# MENSA®

# LOGIC
# PUZZLES

# MENSA® ⊕ ℳ.

# LOGIC PUZZLES

## SHARPEN YOUR WITS AND
## HONE YOUR PROBLEM-SOLVING SKILLS

Philip Carter
Ken Russell

**CARLTON**
BOOKS

# What is Mensa?

Mensa is the international society for people with a high IQ.
We have more than 100,000 members in over 40 countries worldwide.

The society's aims are:
      to identify and foster human intelligence for the benefit of humanity
      to encourage research in the nature, characteristics, and uses of intelligence
      to provide a stimulating intellectual and social environment for its members

Anyone with an IQ score in the top two per cent of the population is eligible
to become a member of Mensa – are you the 'one in 50' we've been looking for?

Mensa membership offers an excellent range of benefits:
      Networking and social activities nationally and around the world
      Special Interest Groups – hundreds of chances to pursue your hobbies
         and interests – from art to zoology!
      Monthly members' magazine and regional newsletters
      Local meetings – from games challenges to food and drink
      National and international weekend gatherings and conferences
      Intellectually stimulating lectures and seminars
      Access to the worldwide SIGHT network for travellers and hosts

For more information about Mensa: www.mensa.org, or

British Mensa Ltd.,
St John's House,
St John's Square,
Wolverhampton
WV2 4AH
Telephone: +44 (0) 1902 772771
E-mail: enquiries@mensa.org.uk
www.mensa.org.uk

# Contents

# Introduction

People often feel that there's something a bit cold about logic. It reminds us of Mr Spock playing multidimensional chess, his mighty Vulcan brain remorselessly analyzing every possible permutation of the game. Even worse, it makes us think of mathematics lessons where we sweated over some ghastly geometrical conundrum that stubbornly refused to yield up its dusty secrets. Fortunately it does not have to be that way.

Ken Russell and Philip Carter, who have for many years been Mensa's resident puzzle experts, have come up with a collection of logic problems just for fun. The nice thing about logic is that it does not require any special knowledge, just a capacity for following an argument one step at a time to its inevitable conclusion. There is something extremely satisfying in being able to take a knotty problem and, after having carefully unpicked the complications, arriving at the solution.

If you like logic problems you will like Mensa, a society that exists entirely for people who are adept at solving the most testing brainteasers. You will find more information on the preceding pages.

*R. P. Allen*

Robert Allen

# Dazzling Diamond

Divide the diamond into four identical shapes, each containing one of each of the following five symbols:

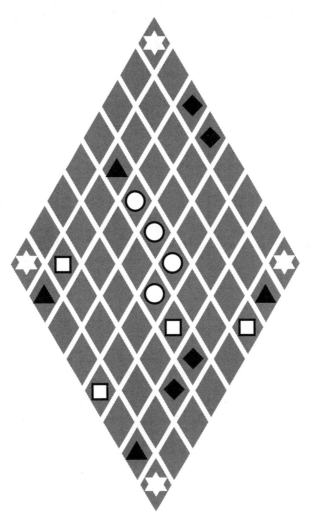

see answer
**21**

# Tricky Triangles

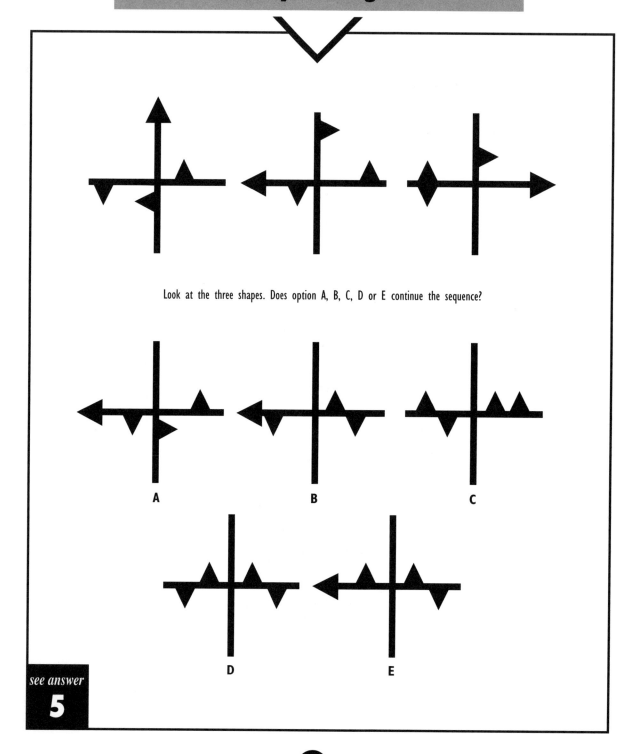

Look at the three shapes. Does option A, B, C, D or E continue the sequence?

A

B

C

D

E

see answer
5

# Roving Robot

Scientists have produced a robot that contains a simple program for crossing a quiet road (not a one-way street) in the UK.

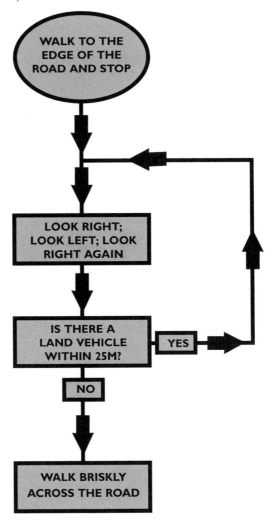

WALK TO THE EDGE OF THE ROAD AND STOP

LOOK RIGHT; LOOK LEFT; LOOK RIGHT AGAIN

IS THERE A LAND VEHICLE WITHIN 25M?

YES

NO

WALK BRISKLY ACROSS THE ROAD

But they made a cardinal error and the robot takes eight hours to cross the road. What is the error?

see answer
20

Draw three complete circles so that each circle contains one ellipse, one square and one triangle and no two circles share exactly the same three elements..

*see answer*
**2**

# Counterfeit Coins

Most counterfeit coin puzzles assume you have balance-type scales available with two pans, where one object is weighed against another. In this puzzle you have a single scale with only one pan. You have three bags of large gold coins with an unspecified number of coins in each bag. One of the bags consists entirely of conterfeit coins weighing 55g each; the other two bags contain all genuine coins weighing 50g each.

What is the minimum number of weighing operations you need to carry out before you can be certain of identifying the bag of counterfeit coins?

see answer
**15**

# Sitting Pretty

At the school the boys sit at desks numbered 1–5 and the girls sit opposite them at desks numbered 6–10.

1. The girl sitting next to the girl opposite no. 1 is Fiona.
2. Fiona is three desks away from Grace.
3. Hilary is opposite Colin.
4. Eddy is opposite the girl next to Hilary.
5. If Colin is not central then Alan is.
6. David is next to Bill.
7. Bill is three desks away from Colin.
8. If Fiona is not central then Indira is.
9. Hilary is three desks away from Jane.
10. David is opposite Grace.
11. The girl sitting next to the girl opposite Alan is Jane.
12. Colin is not at desk no. 5.
13. Jane is not at desk no. 10.

Can you work out the seating arrangements?

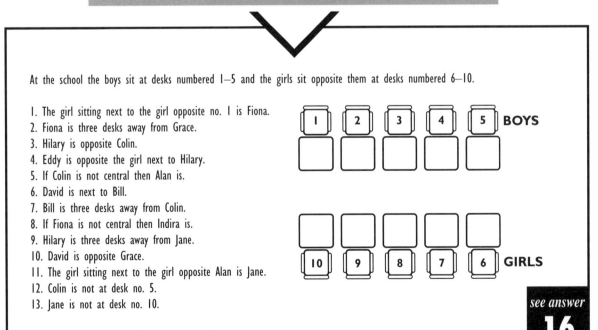

see answer
**16**

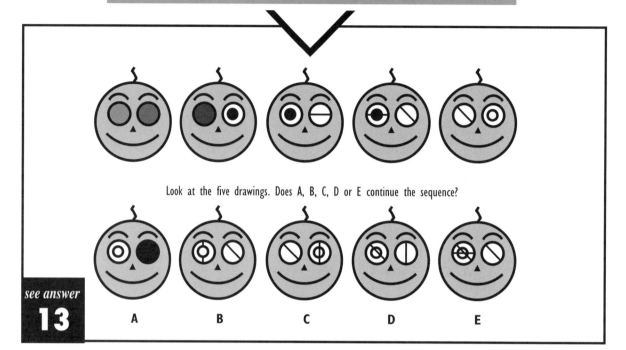

Look at the five drawings. Does A, B, C, D or E continue the sequence?

A     B     C     D     E

see answer
13

# Figure of Fun

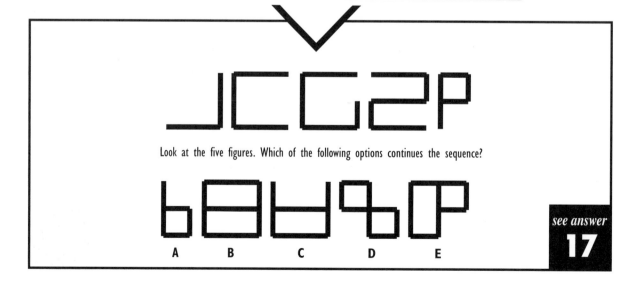

Look at the five figures. Which of the following options continues the sequence?

A     B     C     D     E

see answer
17

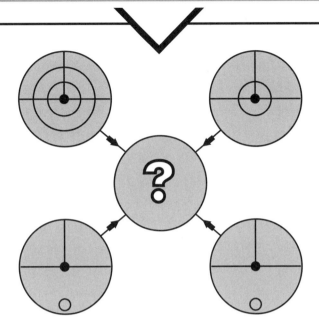

Each line and symbol in the four outer circles is transferred to the middle circle according to a few rules. These are that if a line or symbol occurs in the outer circles:

once, it is transferred; twice, it is possibly transferred; three times, it is transferred; four times, it is not transferred.

Which of the five circles should appear in the middle of the diagram above?

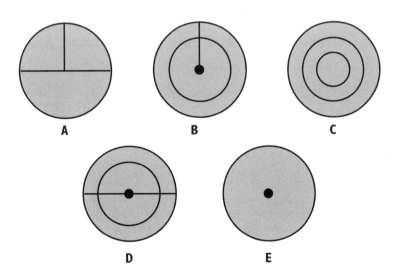

A      B      C

D      E

*see answer*
**10**

# Shooting Range

Three military marksmen — Colonel Present, Major Aim and General Fire — are shooting on the range. When they have finished, they collect their targets, 60 for the inner ring, 40 for the outer ring.

Each makes three statements:

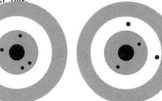

Colonel Present:
"I scored 180."
"I scored 40 less than the major."
"I scored 20 more than the general."

Major Aim:
"I did not score the lowest."
"The difference between my score and the general's was 60."
"The general scored 240."

General Fire:
"I scored less than the colonel."
"The colonel scored 200."
"The major scored 60 more than the colonel."

*see answer*
14

Each marksman makes one incorrect statement. What are their scores?

# Land of Zoz

In the land of Zoz there live three types of person:

Truthkins, who live in hexagonal houses and always tell the truth;
Fibkins, who live in pentagonal houses and always tell lies;
Switchkins, who live in round houses and who make true whatever they say.

One morning 90 of them gather in the city in three groups of 30. One group is all of one type; another group is made up evenly of two types; the third group evenly comprises three types. Everyone in the first group says "We are all truthkins"; everyone in the second group says "We are all fibkins"; and everyone in the third group says "We are all switchkins".

How many sleep in pentagonal houses that night?

*see answer*
12

# Fancy Figures

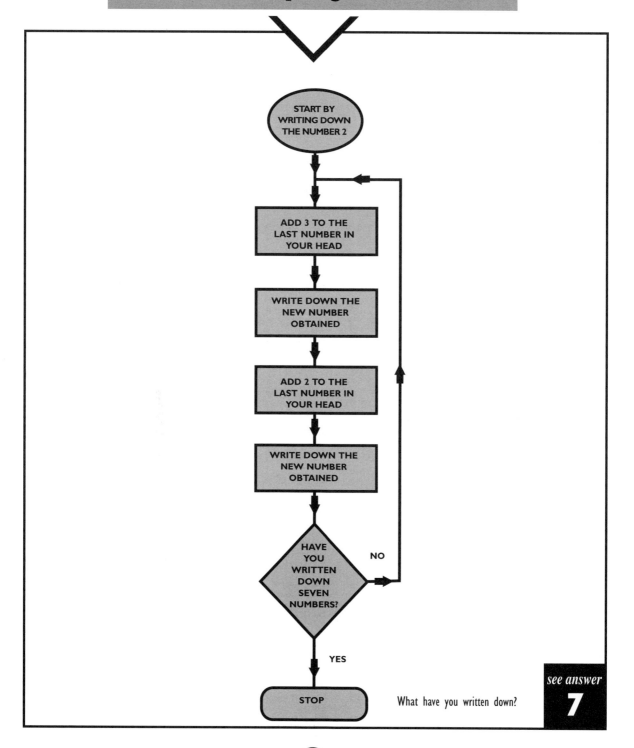

START BY WRITING DOWN THE NUMBER 2

ADD 3 TO THE LAST NUMBER IN YOUR HEAD

WRITE DOWN THE NEW NUMBER OBTAINED

ADD 2 TO THE LAST NUMBER IN YOUR HEAD

WRITE DOWN THE NEW NUMBER OBTAINED

HAVE YOU WRITTEN DOWN SEVEN NUMBERS?

NO

YES

STOP

What have you written down?

*see answer*

**7**

# Booth Bonanza

The new repairer starts work repairing telephones. There are 15 booths in his area. The supervisor tells him that five out of the first eight booths need repairing and that he should go and repair one as a test.

The man goes straight to booth number eight. Why?

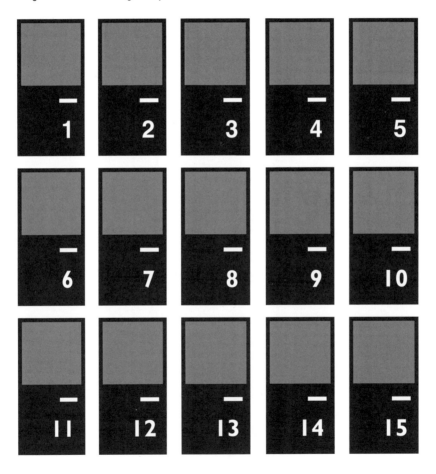

see answer
4

Look at the three hexagons. Which of the following four options continues the sequence?

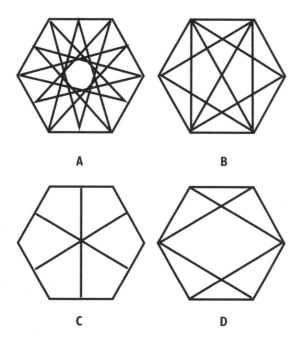

A          B

C          D

*see answer*
**8**

# Las Vegas

Three gamblers — Diablo, Scarface and Lucky — attend a convention at Las Vegas. They decide to have a gambling session with six-sided dice, but stipulate unusual rules:

1. Each gambler may select his own numbers.
2. The numbers 1–9 may be selected, but no two numbers may be consecutive.
3. Each die has to have three pairs of different numbers, adding up to 30.

In addition, no two gamblers are allowed to choose the same two numbers. In a long run, Diablo's numbers will beat Scarface; Scarface's numbers will beat Lucky; but Lucky's numbers will beat Diablo. How is this possible?

**see answer 1**

# Murder in Mind

Four suspects — Jack Vicious, Sid Shifty, Alf Muggins and Jim Pouncer — are being interviewed at the scene of a murder. Each of the suspects is asked a question. Their answers are as follows:

Jack Vicious: "Sid Shifty committed the murder."
Sid Shifty: "Jim Pouncer committed the murder."
Alf Muggins: "I didn't commit the murder."
Jim Pouncer: "Sid Shifty is lying."

Only one of the four answers is the truth. Who committed the murder?

**see answer 6**

# Dynamic Dog

Russell Carter lives on a remote ranch in the Australian outback with his dog, Spot. Several times a week he sets off with Spot for a long walk. This morning he is walking at a steady 4mph and when they are 10 miles from home he turns to go home and, retracing his steps, lets Spot off the lead. Spot immediately runs homeward at 9mph. When Spot reaches the ranch he turns around and runs back to Russell, who is continuing at his steady 4mph. On reaching Russell, Spot turns back for the ranch, maintaining his 9mph. This is repeated until Russell arrives back at the ranch and lets Spot in. At all times Russell and Spot maintain their respective speeds of 4mph and 9mph.

How many miles does Spot cover from being let off the lead to being let into the ranch?

see answer
19

# Manor House

In the English countryside is a traditional manor house. Five staff work there, each of whom has a different hobby and a different rest day.

1. The man who has Tuesday off plays golf but is not the janitor, who is called Clark.
2. Jones is not the butler who plays squash.
3. Wood has Wednesday off and is not the butler or the gardener.
4. James is the cook and does not have Thursdays off; Smith also does not have Thursdays off (or Tuesdays, either).
5. Bridge is played on Monday; the chauffeur does not play chess; and James does not have Tuesdays off.

What are their names, how is each employed, what is the pastime of each, and on which day of the week does each have a rest day?

| | | OCCUPATION | | | | | PASTIME | | | | | DAY | | | | |
|---|---|---|---|---|---|---|---|---|---|---|---|---|---|---|---|---|
| | | BUTLER | CHAUFFEUR | COOK | GARDENER | JANITOR | FISHING | CHESS | SQUASH | BRIDGE | GOLF | MONDAY | TUESDAY | WEDNESDAY | THURSDAY | FRIDAY |
| NAME | SMITH | | | | | | | | | | | | | | | |
| | JONES | | | | | | | | | | | | | | | |
| | WOOD | | | | | | | | | | | | | | | |
| | CLARK | | | | | | | | | | | | | | | |
| | JAMES | | | | | | | | | | | | | | | |
| DAY | MONDAY | | | | | | | | | | | | | | | |
| | TUESDAY | | | | | | | | | | | | | | | |
| | WEDNESDAY | | | | | | | | | | | | | | | |
| | THURSDAY | | | | | | | | | | | | | | | |
| | FRIDAY | | | | | | | | | | | | | | | |
| PASTIME | FISHING | | | | | | | | | | | | | | | |
| | CHESS | | | | | | | | | | | | | | | |
| | SQUASH | | | | | | | | | | | | | | | |
| | BRIDGE | | | | | | | | | | | | | | | |
| | GOLF | | | | | | | | | | | | | | | |

| NAME | OCCUPATION | PASTIME | REST DAY |
|---|---|---|---|
| | | | |
| | | | |
| | | | |
| | | | |
| | | | |

see answer
**9**

# House Hunting

My friend Archibald has moved into a new house in a long road in which the houses are numbered consecutively, 1–82. To find out his house number I ask him three questions to which I receive a yes/no answer. I will not tell you the answers, but if you can work them out you will discover his house number. The questions are:

1. Is it under 41?
2. Is it divisible by 4?
3. Is it a square number?

Can you work out the number of Archibald's house?

*see answer*
**18**

# Town Clock

From my window I can see the town clock. Every day I check the clock on my mantlepiece against the time shown on the town clock. It usually agrees; but one morning a strange situation occurred: on my mantlepiece stands my clock and it showed the time as 5 minutes to 9 o'clock; 1 minute later it read 4 minutes to 9 o'clock; 2 minutes later it read 4 minutes to 9 o'clock; 1 minute later it read 5 minutes to 9 o'clock.

At 9 o'clock I suddenly realized what was wrong. Can you tell what it was?

*see answer*
**3**

# Sears Tower

The national headquarters of Sears Roebuck & Co. in Chicago, Illinois, was, for a while, the tallest inhabited building in the world. Better known as Sears Tower, it is 225m high plus half its height again.

How high is Sears Tower?

see answer

**11**

# Japan Hotel

In a hotel in Nagasaki is a glass door. On the door it says:

# PHUSLULP

What does it mean?

see answer
**22**

# Prisoners' Porridge

A jailer has a large number of prisoners to guard and has to seat them at a number of tables at mealtimes. The regulations state the following seating arrangements:

1. Each table is to seat the same number of prisoners.
2. The number at each table is to be an odd number.

The jailer finds that when he seats the prisoners:
3 per table, he has 2 prisoners left over;
5 per table, he has 4 prisoners left over;
7 per table, he has 6 prisoners left over;
9 per table, he has 8 prisoners left over;
but when he seats them 11 per table there are none left over.

How many prisoners are there?

see answer
**34**

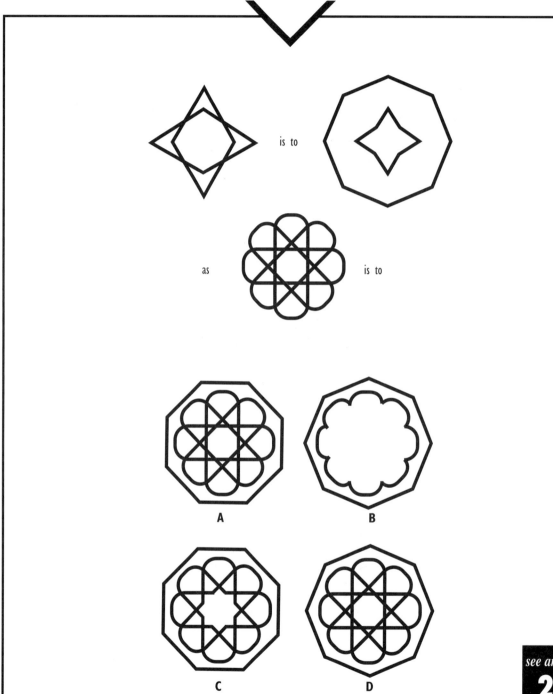

is to

as

is to

A

B

C

D

see answer
25

# Algarve Rendezvous

On the far eastern side of the Algarve, close to the Spanish border, is a town whose roads are laid out in grid fashion, like Manhattan. This system was first used in the cities of Ancient Greece. Seven friends live at different corners, marked . They wish to meet for coffee.

On which corner should they meet in order to minimize the walking distance for all seven?

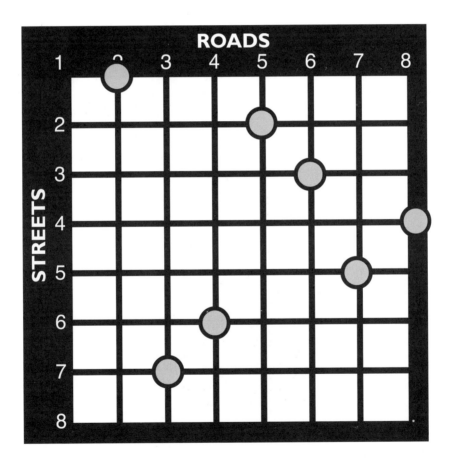

see answer
31

# Intergalactic Ingenuity

Five pairs of husband and wife aliens arrive for the intergalactic meeting on Earth. For ease of recognition, the males are known by the letter M followed by an odd number and the females by F and an even number. Each pair has different distinguishing features and has prepared a different subject for discussion. They arrive in different types of spacecraft and dock in a set of five bays. The pairs sit in five double seats in the auditorium.

1. M1 is preparing his speech on time travel and has arrived in a warp distorter.
2. The mind-reading couple who have four arms each have parked their nebula accelerator between the space oscillator and the astro carrier.
3. F6, in the pair of seats next to the left-end pair, says to the alien next to her, "My husband M3 and I have noticed that you have three legs."
4. F4 admires the galaxy freighter owned by the pair who each have three eyes, who are in the next seats.
5. The husband of F8 is turning his papers on time travel with 12 fingers.
6. M5, in the middle pair of seats, says to F10 in the next pair of seats, "The pair with webbed feet on your other side have an astro carrier."
7. M7 and F2 are studying their papers on anti-gravity. The husband of F6 is studying his papers on nuclear fission.

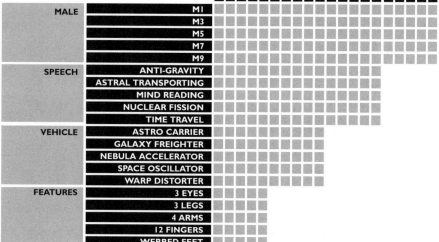

Who is the wife of M9 and who is the male speaker on nuclear fission?

| MALE/FEMALE | | | | | |
|---|---|---|---|---|---|
| SPEECH | | | | | |
| VEHICLE | | | | | |
| FEATURES | | | | | |

see answer
29

# Creative Circles

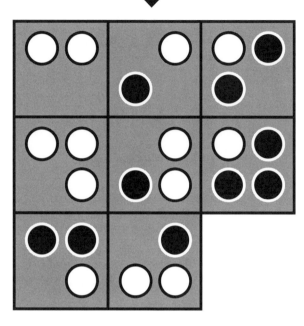

Look along each line and down each column of this shape. Which of the following eight options is the missing square?

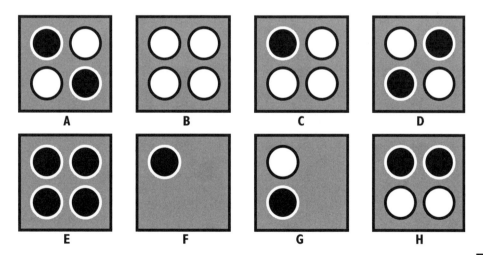

A    B    C    D

E    F    G    H

see answer
35

29

# Searching Segments

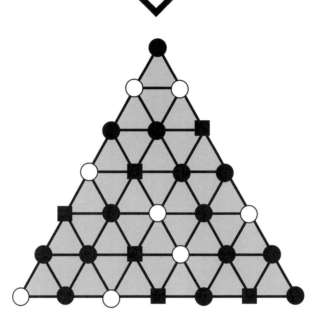

Place the 12 segment links below over the triangular grid in such a way that each link symbol on the grid is covered by an identical symbol. The connecting segments must not be rotated. Not all the connecting lines will be covered.

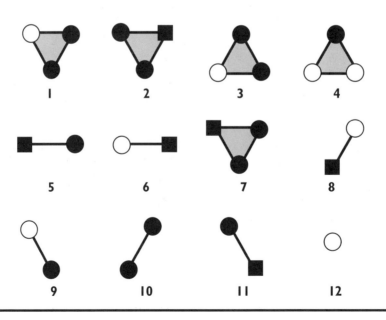

see answer
**32**

30

# Funny Fingers

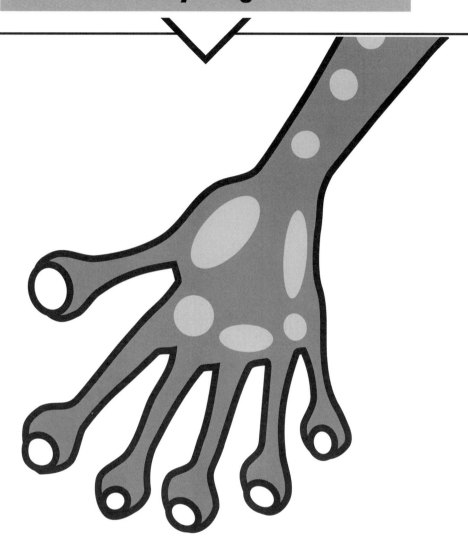

There is a number of aliens in a room, more than one. Each alien has more than one finger on each hand. All aliens have the same number of fingers as each other. All aliens have a different number of fingers on each hand. If you knew the total number of fingers in the room you would know how many aliens were in the room. There are between 200 and 300 alien fingers in the room.

How many aliens are in the room?

see answer
24

# Think of a Number

Anastasia has thought of a number between 99 and 999. Belinda asks whether the number is below 500; Anastasia answers yes. Belinda asks whether the number is a square number; Anastasia answers yes. Belinda asks whether the number is a cube number; Anastasia answers yes. However, Anastasia has told the truth to only two of the three questions. Anastasia then tells Belinda truthfully that both the first and the last digit are 5, 7 or 9.

What is the number?

see answer
**36**

# Girl Talk

A census-taker calls at a house. He asks the woman living there the ages of her three daughters.

The woman says, "If you multiply their ages the total is 72; if you add together their ages the total is the same as the number on my front door, which you can see."

The census-taker says, "That is not enough information for me to calculate their ages."

The woman says, "Well, my eldest daughter has a cat with a wooden leg."

The census-taker replies, "Ah! Now I know their ages."

What are the ages of the three girls?

see answer
**42**

# Treasured Trees

Local sports clubs take turns to plant a tree each year in the town's main street. A bird has established a nest in each tree.

1. The crow lives in the beech tree.
2. The lime was planted two years after the tree planted by the golf club.
3. The robin is in the tree planted by the bowling club, which is next to the tree planted by the soccer club.
4. Jim planted his tree in 1971.
5. The starling is in the poplar tree planted by Desmond in 1974.
6. The robin lives in the tree planted by the bowling club, which is next to the tree planted by the soccer club.
7. Tony planted the middle tree — a beech.
8. Bill has an owl in his tree, which is next to the ash.
9. The tree at the right-hand end was planted in 1974 by the soccer club.
10. The elm was planted in 1970.
11. The tennis club planted in 1972.
12. The squash club planted in 1970.
13. Sylvester planted his tree in 1973 and it has a robin in it.
14. The blackbird is in the tree planted by Jim.

Work out which tree was planted by which member of each club and in which year.

| TREE | | | | | |
|---|---|---|---|---|---|
| PERSON | | | | | |
| CLUB | | | | | |
| BIRD | | | | | |
| YEAR | | | | | |

*see answer*
**28**

## Club Conundrum

There are 189 members of the tennis club: 8 have been at the club less than three years; 11 are under 20 years of age; 70 wear spectacles; 140 are men.

What is the smallest number of players who had been members for three years or more, were at least 20 years of age, wore glasses and were men?

see answer
**38**

## Tree Teaser

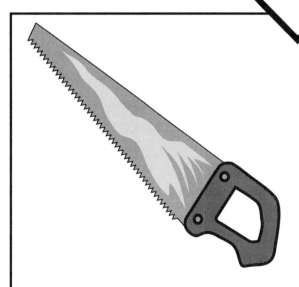

Don and Spencer are engaged by the local council to prune trees on either side of a tree-lined avenue. There is an equal number of trees on either side of the road. Don arrives first and has pruned three trees on the right-hand side when Spencer arrives and points out that Don should be pruning the trees on the left-hand side. So Don starts afresh on the left-hand side and Spencer continues on the right. When Spencer has finished his side he goes across the avenue and prunes six trees for Don, which finishes the job.

Who prunes the most trees and by how many?

see answer
**41**

# Take a Tile

Look at the pattern of tiles. Which of the following tiles replaces the question mark?

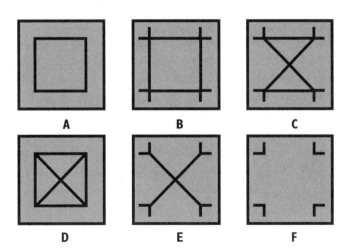

A      B      C

D      E      F

see answer
33

# Dog Delight

There is a somewhat confusing situation at the dog show this year. Four brothers — Andy, Bill, Colin and Donald — each enter two dogs, and each has named his dogs after two of his brothers. Consequently, there are two dogs named Andy, two named Bill, two named Colin and two named Donald.

Of the eight dogs, three are corgis, three labradors and two dalmatians. None of the four brothers owns two dogs of the same breed. No two dogs of the same breed have the same name. Neither of Andy's dogs is named Donald and neither of Colin's dogs is named Andy. No corgi is named Andy and no labrador is named Donald. Bill does not own a labrador.

Who are the owners of the dalmatians and what are the dalmatians' names?

see answer
**37**

# Handkerchief Challenge

Charlie throws out a challenge to Ben in the local bar: "I'll put this ordinary pocket handkerchief on the floor. You stand on one corner and I'll stand on the other corner. Without either of us tearing, cutting, stretching or altering it in any way, I bet you won't be able to touch me."

How can this be done?

see answer
**30**

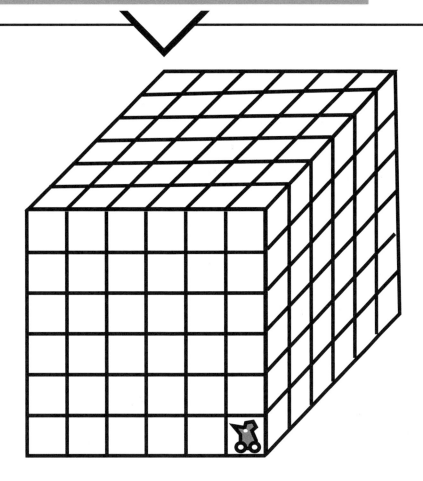

The cage consists of 216 open chambers. An electronic robot
mouse is placed in the bottom right-hand front chamber, marked
above. You are able to operate the mouse by remote control,
moving it three chambers to the right or left and two
chambers up or down.

Are you able to get the mouse into the central chamber, and if
so, what is the minimum number of moves by which
this can be achieved?

*see answer*
**27**

# Roulette Riddle

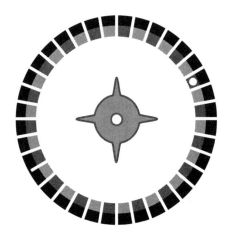

A roulette wheel shows the numbers 1–36. My ball has landed on a particular number. It is divisible by 3. When the digits are added together, the total lies between 4 and 8. It is an odd number. When the digits are multiplied together the total lies between 4 and 8.

Which number have I bet on?

*see answer*
**40**

# Salary Increase

A company gives a choice of two plans to the union negotiator for an increase in salary. The first option is an initial salary of $20,000 to be increased after 12 months by $500. The second option is an initial salary of $20,000 to be increased after each six months by $125. The salary is to be calculated every six months.

Can you advise the union negotiator which is the plan he should recommend to his members?

*see answer*
**23**

# Three Squares

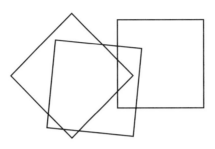

Look at the group of three squares. They have a certain feature which is shared by only one of the groups of three squares below. What is it, and which group matches?

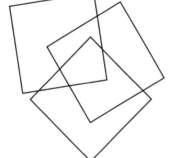

A

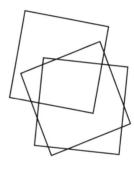

B

C

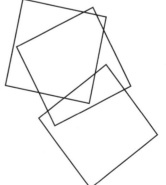

D

*see answer*
**39**

# Barrels of Fun

A wine merchant has six barrels of wine and beer containing

30 gallons
32 gallons
36 gallons
38 gallons
40 gallons
62 gallons

Five barrels are filled with wine and one with beer.
The first customer purchases two barrels of wine; the second customer purchases twice as much wine as the first customer.

Which barrel contains beer?

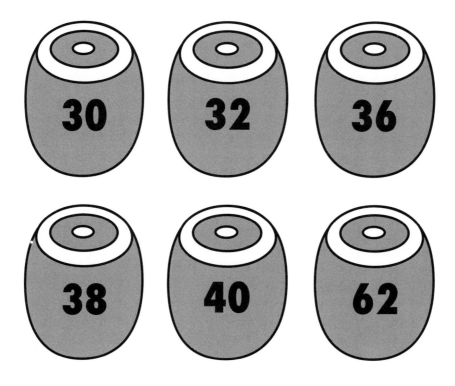

see answer
26

40

## Strange Series

Draw the next figure in this series.

see answer
**58**

## Number Placement

| | 14 | 10 | 7 |
|---|---|---|---|
| 9 | 6 | | 4 |
| 16 | | 13 | 11 |
| 12 | 8 | 5 | 15 |

The numbers 4–16 have already been inserted into the grid, almost — but not quite — at random. Following just two simple rules, where would you place the numbers 1, 2 and 3 in the grid?

see answer
**44**

# Study Time

Three college students — Anne, Bess and Candice — each study four subjects. Two of them study physics; two study algebra; two study English; two study history; two study French; two study Japanese.

Anne: if she studies algebra then she also takes history & French;
if she studies history she does not take English or physics;
if she studies English she does not take Japanese or algebra.

Bess: if she studies English she also takes Japanese and physics;
if she studies Japanese she does not take algebra or history;
if she studies algebra she does not take English or Japanese.

Candice: if she studies French she does not take algebra or English;
if she does not study algebra she studies Japanese and History;
if she studies Japanese she does not take English or physics.

What do you know about these three students?

|  | ANNE | BESS | CANDICE |
|---|---|---|---|
| **PHYSICS** |  |  |  |
| **ALGEBRA** |  |  |  |
| **ENGLISH** |  |  |  |
| **HISTORY** |  |  |  |
| **FRENCH** |  |  |  |
| **JAPANESE** |  |  |  |

see answer 47

# Picking Professions

Mr Carter, Mr Butler, Mr Drover and Mr Hunter are employed as a carter, a butler, a drover and a hunter. None of them has a name identifying their profession. They made four statements:

1. Mr Carter is the hunter.    2. Mr Drover is the carter.    3. Mr Butler is not the hunter.    4. Mr Hunter is not the butler.

According to those statements, the butler must be Mr Butler, but this cannot be correct. Three of the four statements are untrue. Who is the drover?

see answer 49

# SPRING INTO SUMMER!

## JUNE 10TH -12TH

## 15%OFF

REGULAR PRICED BOOKS, GIFTS, TOYS AND MORE*

FIND GREAT GIFTS FOR DAD, A GRAD,
A TEACHER – OR EVEN TREAT YOURSELF!

PRESENT THIS COUPON IN-STORE TO SAVE.

**Chapters**    **!ndigo**    **COLES**

436050206112

# Gone Fishing

Five men staying at the coastal hotel decide to go fishing on the pier. They sit next to each other, using different bait, and catch different numbers of fish.

1. The plumber, called Henry, catches one fish fewer than Dick.
2. The electrician is next to the banker and uses bread for his bait.
3. The man at the north end of the pier is the banker, who is sitting next to Fred.
4. The salesman catches only one fish, and is sitting at the south end of the pier.
5. Meal is the bait used by Malcolm, and the man from Orlando catches 15 fish.
6. The man from New York uses shrimps for bait and is sitting next to the man who catches one fish.
7. Joe is from Los Angeles and uses worms as his bait.
8. The man in the middle is from Tucson and uses a bait of maggots.
9. The banker catches six fish, three less than the man three places away from him.
10. Dick, who is the middle fisherman, is two seats away from the man from St Louis.
11. The man who is sitting next to the man from New York catches 10 fish and is a professor.
12. Henry did not sit next to Joe.

Work out where each man lives, his occupation, the bait he is using, and how many fish he catches.

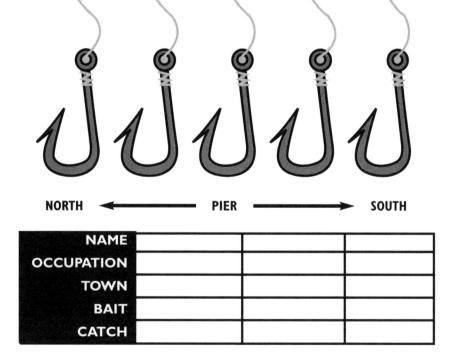

NORTH ⟵ PIER ⟶ SOUTH

| NAME | | | |
|---|---|---|---|
| OCCUPATION | | | |
| TOWN | | | |
| BAIT | | | |
| CATCH | | | |

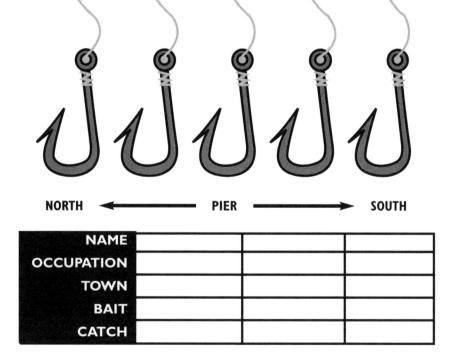

*see answer*

## 54

# Great Golfers

Mr Peters, Mr Edwards and Mr Roberts are playing a round of golf together. Half-way through the game Mr Peters remarks that he has just noticed that their first names are Peter, Edward and Robert. "Yes," says one of the others, "I'd noticed that too, but none of us has the same surname as our own first name. For example, my first name is Robert."

What are the full names of the three golfers?

see answer
**50**

# Best Beer

A man can drink a barrel of beer in 27 days.
A woman can drink a barrel of beer in 54 days.

If they both drink out of the same barrel at their respective rates, how long will it take for the barrel to be emptied?

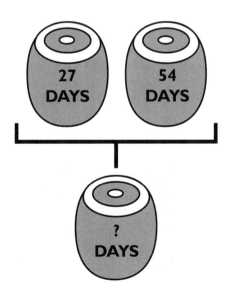

see answer
**63**

# Pleased Pupils

There are five pupils, each in a different class. Each pupil takes a subject and sport which she enjoys.

1. The girl who plays squash likes algebra and is not in class 5.
2. Doris is in class 3 and Betty likes running.
3. The girl who likes running is in class 2.
4. The girl in class 4 likes swimming, while Elizabeth likes chemistry.
5. Alice is in class 6 and likes squash but not geography.
6. The girl who likes chemistry also enjoys basketball.
7. The girl who likes biology also likes running.
8. Clara likes history but not tennis.

Work out the class, subject and sport of each girl.

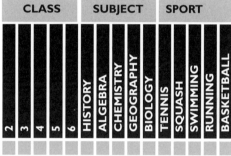

| | | | | | | CLASS | | | | | SUBJECT | | | | | SPORT | | | | |
|---|---|---|---|---|---|---|---|---|---|---|---|---|---|---|---|---|---|---|---|---|
| | | | | | | 2 | 3 | 4 | 5 | 6 | HISTORY | ALGEBRA | CHEMISTRY | GEOGRAPHY | BIOLOGY | TENNIS | SQUASH | SWIMMING | RUNNING | BASKETBALL |
| **NAME** | ALICE | | | | | | | | | | | | | | | | | | | |
| | BETTY | | | | | | | | | | | | | | | | | | | |
| | CLARA | | | | | | | | | | | | | | | | | | | |
| | DORIS | | | | | | | | | | | | | | | | | | | |
| | ELIZABETH | | | | | | | | | | | | | | | | | | | |
| **SPORT** | TENNIS | | | | | | | | | | | | | | | | | | | |
| | SQUASH | | | | | | | | | | | | | | | | | | | |
| | SWIMMING | | | | | | | | | | | | | | | | | | | |
| | RUNNING | | | | | | | | | | | | | | | | | | | |
| | BASKETBALL | | | | | | | | | | | | | | | | | | | |
| **SUBJECT** | HISTORY | | | | | | | | | | | | | | | | | | | |
| | ALGEBRA | | | | | | | | | | | | | | | | | | | |
| | CHEMISTRY | | | | | | | | | | | | | | | | | | | |
| | GEOGRAPHY | | | | | | | | | | | | | | | | | | | |
| | BIOLOGY | | | | | | | | | | | | | | | | | | | |

| NAME | CLASS | SUBJECT | SPORT |
|---|---|---|---|
| | | | |
| | | | |
| | | | |
| | | | |
| | | | |

see answer
**52**

# Winning Wager

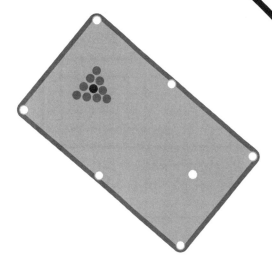

Bill says to Jim, "Let's have a wager on each frame. We will play for half of the money in your wallet on each frame, and we will have 10 frames. Since you have $8 in you wallet, we will play for $4 on the first frame. I will give you $4 if you win and you will give me $4 if I win. When we start the second frame you will have either $12 or $4, so we will play for $6 or $2, etc."

They play 10 frames. Bill wins four and loses six frames but Jim finds that he has only $5.70 left and so has lost $2.30. How is this possible?

see answer
**55**

# Cube Diagonals

Two diagonals have been drawn on two faces of the cube. Using logical reasoning and lateral thinking, can you work out the angle between the two diagonals AB and AC?

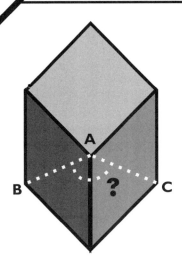

see answer
**62**

## Lost Time

A clock on the wall falls to the floor and the face breaks into three pieces. The digits on each piece of clock add up to the same total. What are the digits on each piece?

*see answer*
**51**

## Changing Trains

A woman usually leaves work at 5.30pm, calls at the supermarket, then catches the 6pm train, which arrives at the station in her home town at 6.30pm. Her husband leaves home each day, drives to the station and picks her up at 6.30pm, just as she gets off the train.

Today the woman finishes work about five minutes earlier than usual, decides to go straight to the station instead of calling at the supermarket, and manages to catch the 5.30pm train, which arrives at her home station at 6pm. Since her husband is not there to pick her up she begins to walk home. Her husband leaves home at the usual time, sees his wife walking, turns around, picks her up and drives home, arriving there 10 minutes earlier than usual.

Assume that all the trains arrive precisely on time. For how long does the woman walk before her husband picks her up?

*see answer*
**59**

# Spot the Shape

Each of the nine squares in the grid marked 1A to 3C should incorporate all the lines and symbols which are shown in the squares of the same letter and number immediately above and to the left. For example, 2B should incorporate all the lines and symbols that are in 2 and B.

One of the squares is incorrect. Which is it?

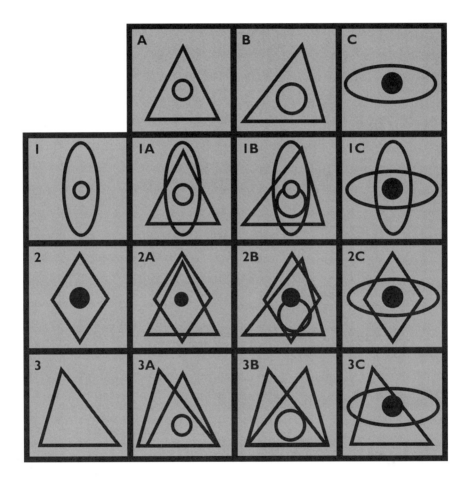

see answer
57

# Fathers and Daughters

A group of friends get together with their daughters for the evening.

1. John is 52 years old and his daughter is not called Eve.
2. Len has a daughter aged 21 years, and Betty is three years older than Eve.
3. Kevin is 53 years old and Diana is 19 years old.
4. Eve is 18 years old, and Nick has a daughter called Carol.
5. Alison is 20 years old and her father is called John.
6. Kevin has a daughter aged 19 years, and Eve's father is called Malcolm.
7. Malcolm is three years older than Nick.

| | | FATHER'S AGE | | | | | DAUGHTER'S AGE | | | | | DAUGHTER | | | | |
|---|---|---|---|---|---|---|---|---|---|---|---|---|---|---|---|---|
| | | 50 | 51 | 52 | 53 | 54 | 17 | 18 | 19 | 20 | 21 | ALISON | BETTY | CAROL | DIANA | EVE |
| FATHER | JOHN | | | | | | | | | | | | | | | |
| | KEVIN | | | | | | | | | | | | | | | |
| | LEN | | | | | | | | | | | | | | | |
| | MALCOLM | | | | | | | | | | | | | | | |
| | NICK | | | | | | | | | | | | | | | |
| DAUGHTER | ALISON | | | | | | | | | | | | | | | |
| | BETTY | | | | | | | | | | | | | | | |
| | CAROL | | | | | | | | | | | | | | | |
| | DIANA | | | | | | | | | | | | | | | |
| | EVE | | | | | | | | | | | | | | | |
| DAUGHTER'S AGE | 17 | | | | | | | | | | | | | | | |
| | 18 | | | | | | | | | | | | | | | |
| | 19 | | | | | | | | | | | | | | | |
| | 20 | | | | | | | | | | | | | | | |
| | 21 | | | | | | | | | | | | | | | |

| FATHER | DAUGHTER | FATHER'S AGE | DAUGHTER'S AGE |
|---|---|---|---|
| | | | |
| | | | |
| | | | |
| | | | |
| | | | |

see answer 60

A woman has two sons, Graham and Frederick. Frederick is three times as old as Graham. If you square Frederick's age you arrive at the same total as when you cube Graham's age. If you subtract Graham's age from Frederick's you arrive at the number of steps in the path to the family's front door. If you add Graham's age to Frederick's you arrive at the number of palisades in the family's fence. If you multiply their ages you arrive at the number of bricks in the family's front wall.

If you add these last three numbers together you have the family's house number, which is 297.

How old are Graham and Frederick?

see answer
## 61

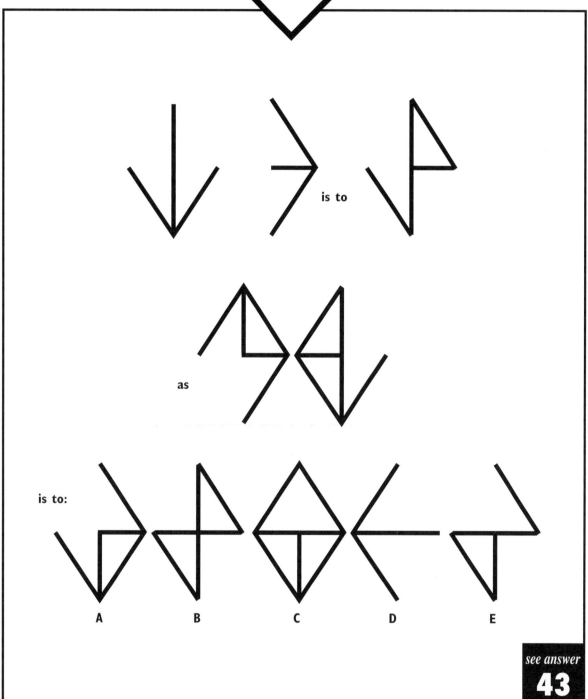

is to

as

is to:

A       B       C       D       E

see answer
**43**

# Play Watching

Four husband and wife couples go to see a play. They all sit in the same row, but no husband sits next to his wife, and a man and a woman are at opposite ends of the row. Their names are Andrews, Barker, Collins and Dunlop.

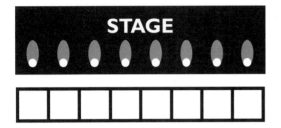

1. Mrs Dunlop or Mr Andrews is in the end seat.
2. Mr Andrews is mid-way between Mr Collins and Mrs Collins.
3. Mr Collins is two seats from Mrs Dunlop.
4. Mrs Collins is mid-way between Mr and Mrs Barker.
5. Mrs Andrews is next to the end seat.
6. Mr Dunlop is two seats from Mr Andrews.
7. Mrs Collins is closer to the right end than the left end.

Work out the seating arrangements along the row.

Feed information into the seats

|   | | | | | | | | |
|---|---|---|---|---|---|---|---|---|
| 1 | | | | | | | | |
| 2 | | | | | | | | |
| 3 | | | | | | | | |
| 4 | | | | | | | | |
| 5 | | | | | | | | |
| 6 | | | | | | | | |
| 7 | | | | | | | | |

see answer
45

# Trying Triangles

What is the smallest number of segments of equal area and
shape that the rectangle can be divided into so that each
segment contains the same number of triangles?

see answer
**48**

# Hello Sailor

Five sailors of different rank are at different ports on different ships.

1. Manning is at the Falklands, and the purser is Dewhurst.
2. Brand is on a warship, and the purser is not on the cruiser.
3. Perkins is on the aircraft carrier, and Ward is at Portsmouth.
4. The commander is at the Falklands, and Manning is on a submarine.
5. The warship is at Crete, and Perkins is at Malta.
6. The frigate is at Gibraltar and the steward is at Malta.
7. Brand is a captain and the seaman is not on the frigate.

Work out the details of each sailor.

| | | RANK | | | | | SHIP | | | | | LOCATION | | | |
| --- | --- | --- | --- | --- | --- | --- | --- | --- | --- | --- | --- | --- | --- | --- | --- |
| | | COMMANDER | CAPTAIN | STEWARD | PURSER | SEAMAN | CRUISER | WARSHIP | FRIGATE | SUBMARINE | AIRCRAFT CARRIER | MALTA | CRETE | FALKLANDS | GIBRALTAR | PORTSMOUTH |
| **NAME** | PERKINS | | | | | | | | | | | | | | | |
| | WARD | | | | | | | | | | | | | | | |
| | MANNING | | | | | | | | | | | | | | | |
| | DEWHURST | | | | | | | | | | | | | | | |
| | BRAND | | | | | | | | | | | | | | | |
| **LOCATION** | MALTA | | | | | | | | | | | | | | | |
| | CRETE | | | | | | | | | | | | | | | |
| | FALKLANDS | | | | | | | | | | | | | | | |
| | GIBRALTAR | | | | | | | | | | | | | | | |
| | PORTSMOUTH | | | | | | | | | | | | | | | |
| **SHIP** | CRUISER | | | | | | | | | | | | | | | |
| | WARSHIP | | | | | | | | | | | | | | | |
| | FRIGATE | | | | | | | | | | | | | | | |
| | SUBMARINE | | | | | | | | | | | | | | | |
| | AIRCRAFT CARRIER | | | | | | | | | | | | | | | |

| NAME | RANK | SHIP | LOCATION |
| --- | --- | --- | --- |
| | | | |
| | | | |
| | | | |
| | | | |
| | | | |

see answer
**53**

Which of these circles is the odd one out?

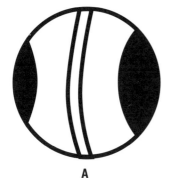

**A**

**B**

**C**

**D**

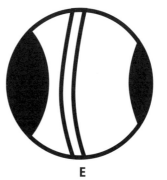

**E**

see answer
**46**

# Odd One Out

Which of the following six shapes is the odd one out?

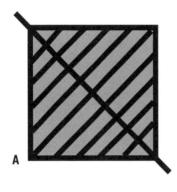

A

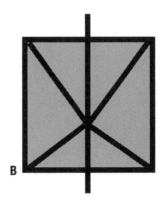

B

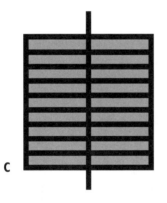

C

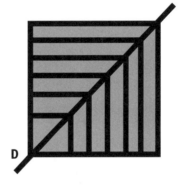

D

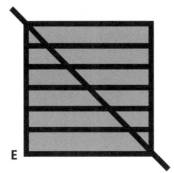

E

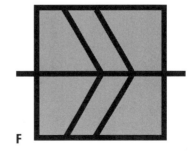

F

see answer
56

56

## Broadway, NY

In Broadway, New York City, a man sees one of those new-fangled buses. It was stationary and he could not tell which way it was going.

Can you?

see answer
**74**

A ⟵        B ⟶

## Logical Clocks

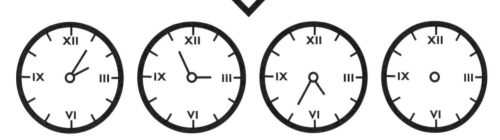

These clocks follow a weird kind of logic. What time should the fourth clock show? Choose from the four options provided.

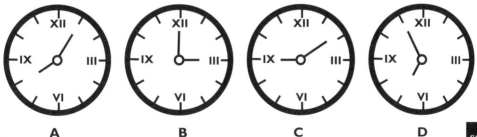

A          B          C          D

see answer
**73**

# Round and Round the Garden

A woman has a garden path 2m wide, demarcated with pebbles, which spiral tightly into the middle of the garden. One day the woman walks the length of the path, finishing in the middle. Ignore the width of the hedge and assume she walks in the middle of the path. How far does she walk?

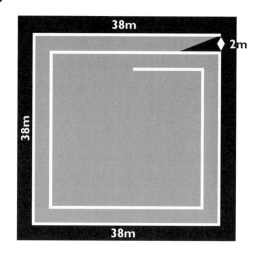

see answer
80

# Rifle Range

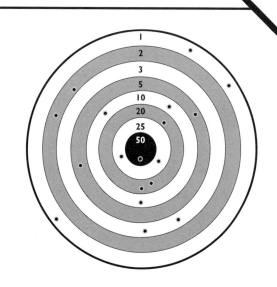

Three soldiers — Colonel Ketchup, Major Mustard and Captain Chutney — have a shooting competition. They each fire six shots and each score 71 points. Colonel Ketchup's first two shots score 22; Major Mustard's first shot scores 3.

Who hits the bull's eye?

see answer
67

# Missing Links

| 74882 | 3584 | |
|---|---|---|

| 29637 | | 192 |
|---|---|---|

| 74826 | | |
|---|---|---|

Fill in the missing numbers from these bars. Just enough information has been provided to work out the logic. The logic is the same in each line of numbers.

Now try this one:

| 528 | 116 | |
|---|---|---|

| 793 | | 335 |
|---|---|---|

| 821 | | |
|---|---|---|

*see answer*
**76**

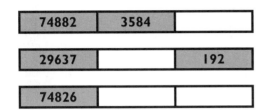

# Gun Running

Bully Bill and Dynamo Dan are cattle ranchers. One day they decide to sell their stock and become sheep farmers. They take the cattle to market and receive for each steer a number of dollars equal to the total number of steer that they sell. With this money they purchase sheep at $10 per head, and with the money left over they purchase a goat.

On the way home they argue and so decide to divide up their stock, but find that they have one sheep over. So Bully Bill keeps the sheep and gives Dynamo Dan the goat.

"But I have less than you," says Dynamo Dan, "because a goat is worth less than a sheep."

"Alright," says Bully Bill, "I will give you my Colt .45 to make up the difference."

What is the value of the Colt .45?

*see answer*
**65**

# Black and White Balls

This probability problem can be solved through logical thought.

You have two bags, each one containing eight balls: four white and four black. A ball is drawn out of bag one and another ball out of bag two.

What are the chances that at least one of the balls is black?

**BAG ONE**

**BAG TWO**

*see answer*
**70**

# Triangles and Trapeziums

is to

as

is to

A    B    C    D

*see answer*
**83**

# Dice Dilemma

Here are views of six non-standard, six-sided dice. Which of the six dice cannot be made up from the following?

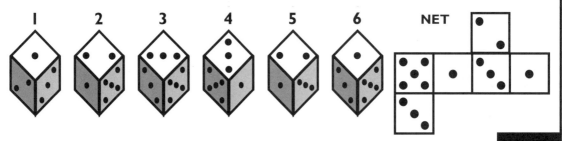

1    2    3    4    5    6    NET

see answer
**82**

# Carrier Pigeons

A driver approaches a bridge. He notices that the maximum weight allowed is 20 tons and knows that his empty pantechnicon weighs 20 tons. However, he has a cargo of 200 pigeons which weigh 1lb each. As the pigeons are asleep on perches he stops the vehicle, bangs on the side to waken the birds who start flying around, then safely drives over the bridge.

Is he correct?

BRIDGE
MAX. WEIGHT
20 TONS

HOMING
PIGEONS

see answer
**71**

# Circles and Triangles

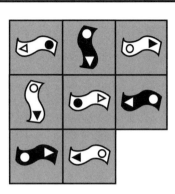

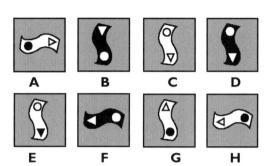

Look along each line and down each column and work out which is the missing square from these options:

see answer
**64**

# City Slicker

City blocks have been built between two main roads — A and B — in a grid, like Manhattan, New York. Always moving toward B, how many different routes are there?

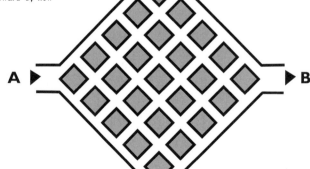

see answer
**77**

# Missing Numbers

There is a logic behind the distribution of numbers in the grid.
Work out what it is and then fill in the missing numbers.

| 6 | 4 | 7 | 8 | 3 | 7 |
|---|---|---|---|---|---|
| 8 | 2 | 5 | 1 | 5 | 6 |
| 3 |   | 8 | 6 | 4 | 8 |
| 8 | 6 | 5 | 3 | 7 | 6 |
| 5 | 4 | 7 |   |   | 5 |
|   | 8 | 6 | 4 | 7 | 8 |

see answer
**66**

# Door Number Puzzle

Two workmen are putting the finishing touches to a new door they have fitted to house number 4761. All that is left to do is to screw the four metal digits to the door. Being a Mensan, Patrick could not resist challenging Bruce by asking him if he could screw the digits onto the door to give a four-figure number which could not be divided exactly by 9. When they had resolved that puzzle Bruce then asked Patrick if he could screw the same digits onto the door to give a four-figure number which could not be divided exactly by 3.

What are the answers to the two puzzles? Can either of them be done?

see answer
**81**

# Star Gazing

What is the largest star that can be drawn so that it is in the same proportions as the other stars and so that it does not touch another star or overlap the edges of the border?

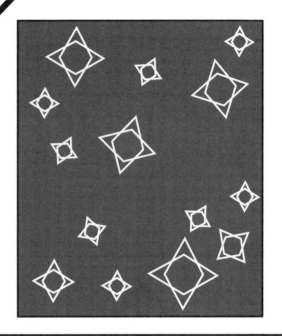

see answer
**78**

# Skyscraper Sizzler

A woman lives in a skyscraper 36 floors high and served by several elevators, which stop at each floor going up and down. Each morning she leaves her apartment on one of the floors and goes to one of the elevators. Whichever one she takes is three times more likely to be going up than down.

Why is this?

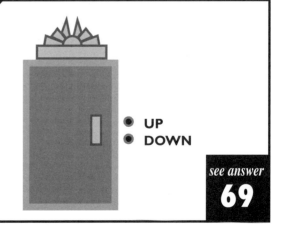

UP
DOWN

see answer
**69**

# Round the Hexagons

Can you work out what should be the contents of the top hexagon?

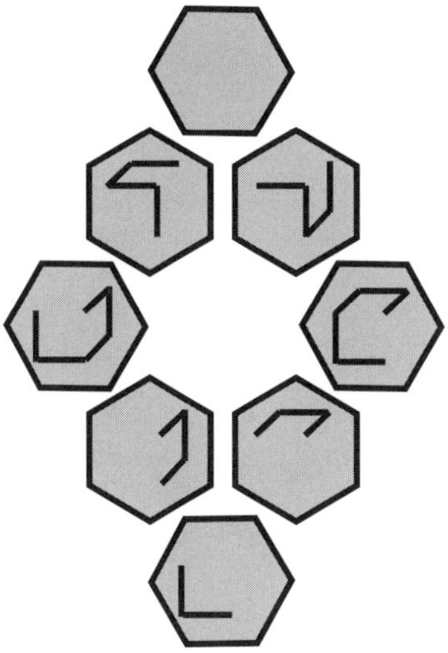

Choose from option

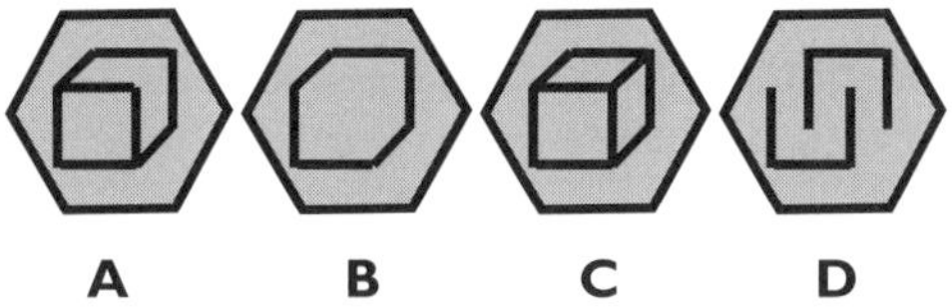

A    B    C    D

see answer
68

# Water Divining

Two men are arguing about whether a square open-topped water tank is half full or not. How can they decide without removing the water or using any measuring device?

see answer
75

# Bartender's Beer

A man goes into a bar in New York. "Glass of beer" he says to the bartender. "Light or special?" asks the bartender. "What's the difference?" asks the man. "Light is 90 cents, special is $1," replies the bartender. "I'll have the special," says the man, placing $1 on the counter.

Another man comes into the bar. "Glass of beer please," he says, placing $1 on the counter. The bartender gives him the special.

Why does he do that?

see answer
**72**

# Pyramid Puzzle

Look at the sequence of shapes.
Which of the following options carries on the sequence?

A      B      C      D      E

see answer
**79**

# Ski-lift

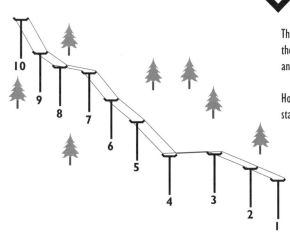

There are 10 places to embark and disembark on the ski-lift at the ski resort. It is possible to purchase a single ticket between any two stations.

How many different tickets are needed for skiers to go to every station from every other station?

see answer
84

# Round in Circles

Look at the four circles. Which of the following circles comes next in the sequence?

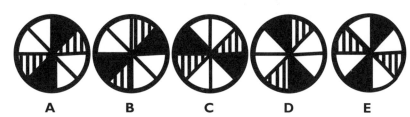

A    B    C    D    E

see answer
110

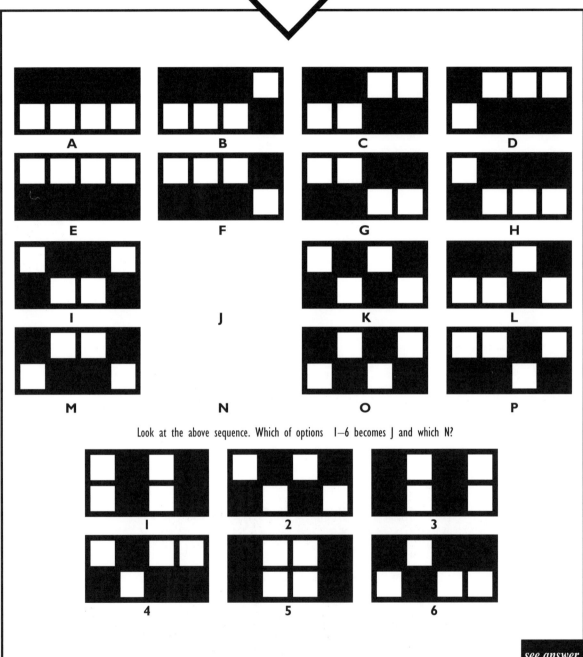

Look at the above sequence. Which of options 1–6 becomes J and which N?

see answer
**86**

# Pyramidal Logic

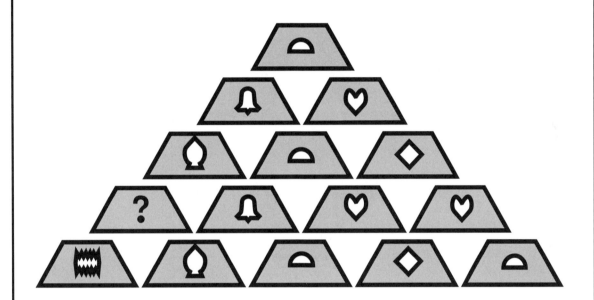

Look at the pyramid. Which of the following symbols should replace the question mark?

A  B  C  D  E

see answer
89

# Pentagon Figures

What number should replace the question mark?

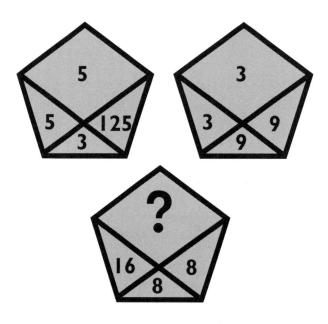

see answer
101

# Sun Shine

There is a valley somewhere on the Earth. The Sun is nearer the valley by over 4,800km at noon than it is when it rises or sets.

Where is this valley?

see answer
91

Which of the following figures is the odd one out?

**A**

**B**

**C**

**D**

**E**

*see answer*
**96**

At the fairground there is a competition — you purchase a ticket on which there is a number of scratch-off squares. One square is marked "loser"; two others have identical symbols. If these appear before the loser square appears, you win a prize. The odds against winning are 2:1 against.

How many squares are there on the card?

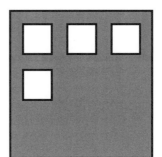

*see answer*
**97**

What number should replace the question mark?

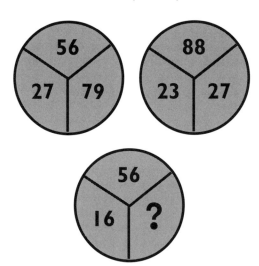

see answer
## 94

# Dinner Party Placements

Mr and Mrs Ackrington, Mr and Mrs Blackpool, Mr and Mrs Chester and Mr and Mrs Doncaster are attending a dinner party. Only one couple does not sit next to each other, and this couple does not sit across from each other. The person sitting opposite Mrs Accrington is a man who is sitting immediately to the left of Mr Blackpool. The person sitting on Mrs Chester's immediate left is a man who is sitting across from Mr Doncaster.

Which couple does not sit next to each other?

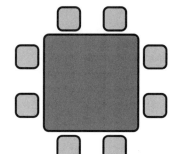

see answer
## 87

# Square Sort

This grid consists of three squares marked A, B and C, and three squares marked 1, 2 and 3. The nine inner squares should incorporate the lines and symbols of both the letter and the number squares. One of the nine squares is incorrect. Which is it?

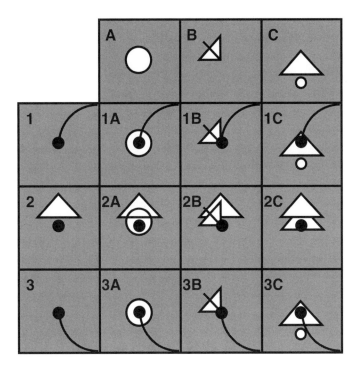

see answer
109

# Generation Gap

I am four times as old as my daughter. In 20 years time I shall be twice as old as her. How old are we today?

see answer
93

# Five Circles

All five circles have the same diameter. Draw a line moving through point A in such a way that it divides the five circles into two equal areas.

see answer
**105**

# Lateral Logic

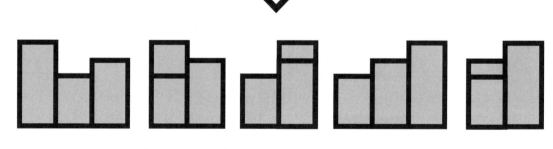

Look at these shapes. Does option A, B, C, D or E continue the sequence?

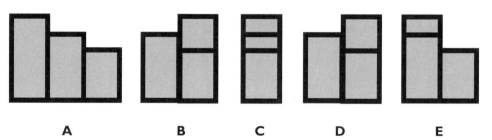

A        B        C        D        E

see answer
**99**

Look at the three squares. Does option A, B, C, D, E or F continue the sequence?

**A**      **B**      **C**      **D**

**E**      **F**

see answer
92

# Figure Columns

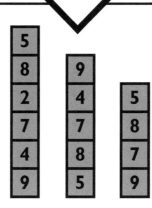

Look at the three columns of figures. Which column comes next in the sequence?

| A | B | C | D |
|---|---|---|---|
| 7 | 8 | 9 | 9 |
| 8 | 7 | 8 | 7 |
| 5 | 9 | 5 | 8 |

see answer
**90**

# Eighteen Trees

A gardener has 18 trees which he wishes to plant in straight rows of five trees per row. He sets himself the task of planting the 18 trees in such an arrangement that he will obtain the maximum number of rows of five trees per row.

There are two slightly different ways he can do this. Can you find both ways?

see answer
**103**

# Notable Number

Fill in the missing number.

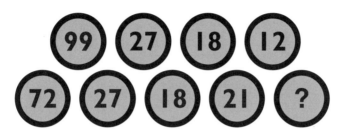

see answer
**100**

# Counting Creatures

At the zoo there are penguins and huskies next to each other. In all, I can count 72 creatures and 200 legs.

How many penguins are there?

see answer
**102**

# Line Analogy

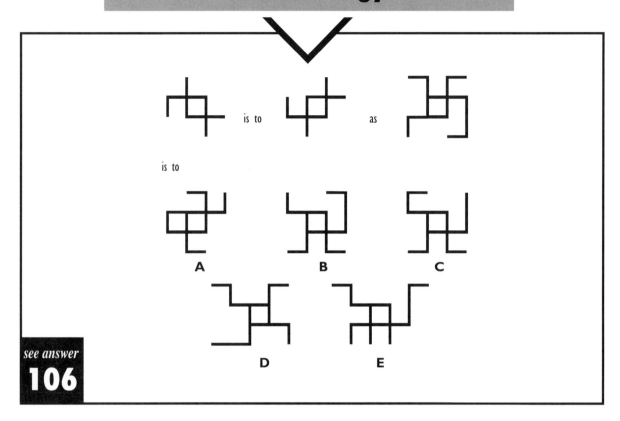

is to ... is to

A    B    C

D    E

see answer
106

# Chess Strategy

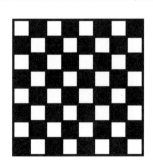

A man has to win two games of chess in a row in order to win a prize. In total, he has to play only three games, alternating between a strong and a weak opponent. Should he play strong, weak, strong; or weak, strong, weak?

see answer
95

## Careful Calculation

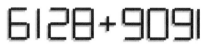

**6128+9091**

If the two numbers total 9825, what do the two numbers below total?

**8159+1912**

see answer
88

## Number Crunching

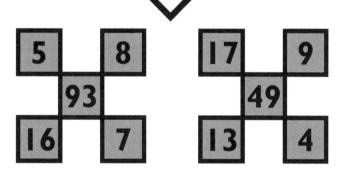

| 5 | | 8 |
|---|---|---|
| | 93 | |
| 16 | | 7 |

| 17 | | 9 |
|---|---|---|
| | 49 | |
| 13 | | 4 |

Look at the diagrams. What number should replace the question mark?

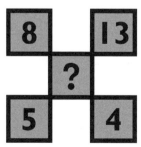

| 8 | | 13 |
|---|---|---|
| | ? | |
| 5 | | 4 |

see answer
108

# Five Pilots

Five pilots take their flights from five different UK airports to five different countries.

Can you sort them out?

1. The aircraft from Stansted flies to Nice.
2. The flight from Cardiff has a captain named Paul.
3. Mike flies to JFK, New York, but not from Gatwick.
4. The flight from Manchester does not go to the USA.
5. Nick flies to Vancouver.
6. Paul does not fly to Roma.
7. Nick does not fly from Manchester.
8. Robin does not fly from Stansted.
9. The flight from Heathrow, not piloted by Tony, is not for Berlin.

| NAME | AIRPORT | DESTINATION |
|------|---------|-------------|
|      |         |             |
|      |         |             |
|      |         |             |
|      |         |             |
|      |         |             |

see answer
85

# Frogs and Flies

If 29 frogs catch 29 flies in 29 minutes, how many frogs are required to catch 87 flies in 87 minutes?

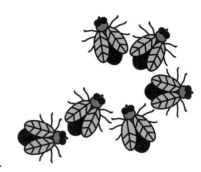

see answer
**98**

# Island Access

There is a lake with an island in the middle. On the island is a tree. The lake is deep and is 80 yards in diameter. There is another tree on the mainland. A non-swimmer wishes to get accross to the island, but all he has is a length of rope 300 yards long.

How does he get accross?

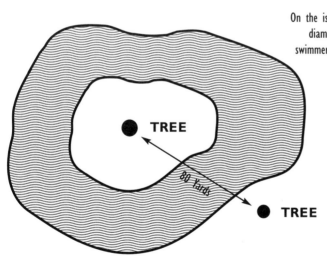

TREE

80 Yards

TREE

see answer
**107**

# Fairground Fiesta

At the carnival five boys of different ages eat different foods and take different rides.

1. Ron eats ice cream, Joe does not chew gum.
2. Sam, who is 14 years old, is not on the mountain.
3. The boy on the crocodile is 15 years old.
4. Len is not on the dodgems; Don is on the whirligig.
5. The boy eating ice cream is 13 years old.
6. The boy on the dodgems is eating a hot dog.
7. Joe eats fries on the big dipper.
8. Don, who is 12, is eating candy floss.

Work out the details of each boy.

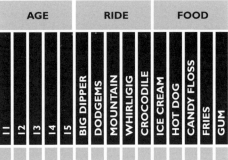

| NAME | AGE | RIDE | FOOD |
|------|-----|------|------|
|  |  |  |  |
|  |  |  |  |
|  |  |  |  |
|  |  |  |  |
|  |  |  |  |

see answer 104

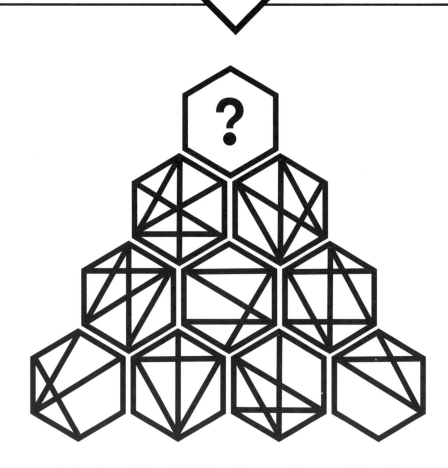

Look at the pyramid. From the following options, choose the contents of the top hexagon.

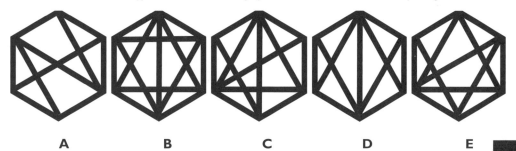

A          B          C          D          E

*see answer*
**116**

# Logic Circles

Look at the four circles. Should A, B, C, D or E follow on the sequence?

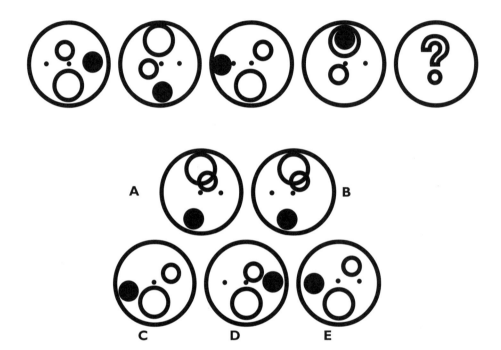

*see answer*
**111**

# Suspicious Shape

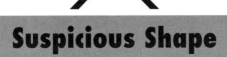

Which of these shapes is the odd one out?

A  B

C D E

F G

*see answer*
**113**

# Sequence Search

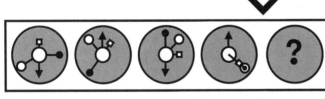

Look at the four circles. Which of the following options comes next in this sequence?

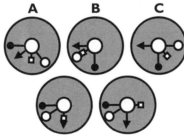

A    B    C

D    E

*see answer*
**117**

# Bird Fanciers

Five men from different countries each like a different bird. Each bird has a different collective noun.

1. Roger does not like plovers, which are not called a parliament.
2. The man who likes crows comes from France. This is not Edward, who is not from Scotland.
3. Albert likes owls; a group of starlings is called a murmuration.
4. Harold comes from Germany and likes ravens.
5. The man from England likes starlings.
6. Edward does not like the group called an unkindness of ravens.
7. The man who likes the group called murder comes from France.
8. Cameron is not from Belgium; Albert is not from Scotland.
9. The man who likes the group called wing is not from Germany.

Work out the details for each man.

|  |  | COUNTRY | | | | | BIRD | | | | | COLLECTIVE NOUN | | | | |
|---|---|---|---|---|---|---|---|---|---|---|---|---|---|---|---|---|
|  |  | GERMANY | BELGIUM | FRANCE | ENGLAND | SCOTLAND | OWLS | PLOVERS | STARLINGS | CROWS | RAVENS | MURMURATION | WING | UNKINDNESS | MURDER | PARLIAMENT |
| NAME | ALBERT |  |  |  |  |  |  |  |  |  |  |  |  |  |  |  |
|  | ROGER |  |  |  |  |  |  |  |  |  |  |  |  |  |  |  |
|  | HAROLD |  |  |  |  |  |  |  |  |  |  |  |  |  |  |  |
|  | CAMERON |  |  |  |  |  |  |  |  |  |  |  |  |  |  |  |
|  | EDWARD |  |  |  |  |  |  |  |  |  |  |  |  |  |  |  |
| COLLECTIVE NOUN | MURMURATION |  |  |  |  |  |  |  |  |  |  | | | | | |
|  | WING |  |  |  |  |  |  |  |  |  |  | | | | | |
|  | UNKINDNESS |  |  |  |  |  |  |  |  |  |  | | | | | |
|  | MURDER |  |  |  |  |  |  |  |  |  |  | | | | | |
|  | PARLIAMENT |  |  |  |  |  |  |  |  |  |  | | | | | |
| BIRD | OWLS |  |  |  |  |  | | | | | | | | | | |
|  | PLOVERS |  |  |  |  |  | | | | | | | | | | |
|  | STARLINGS |  |  |  |  |  | | | | | | | | | | |
|  | CROWS |  |  |  |  |  | | | | | | | | | | |
|  | RAVENS |  |  |  |  |  | | | | | | | | | | |

| NAME | COUNTRY | BIRD | COLLECTIVE NOUN |
|---|---|---|---|
|  |  |  |  |
|  |  |  |  |
|  |  |  |  |
|  |  |  |  |
|  |  |  |  |

see answer
114

# Symbol Search

Each line and symbol which appears in the four outer circles is transferred to the middle circle according to a few rules. If a line or symbol occurs in the outer circles:

once, it is transferred;

twice, it is possibly transferred;

three times, it is transferred;

four times, it is not transferred.

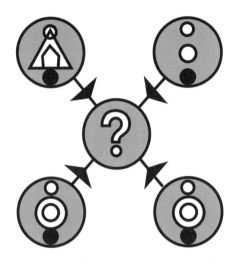

Which of the following circles should appear at the centre of the diagram?

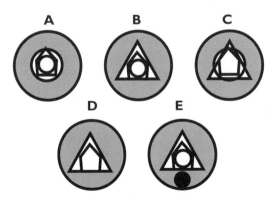

see answer
## 129

Correct this equation so that it makes sense by freely moving the given four digits but without introducing any additional mathematical symbols.

*see answer*
**112**

## Trying Trominoes

Consider the three trominoes. Now choose one of the following to accompany them.

A     B     C     D     E

*see answer*
**119**

# Unwanted Guest

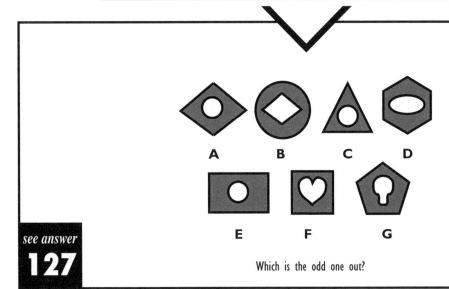

A  B  C  D

E  F  G

see answer
127

Which is the odd one out?

# Round the Circle

What is the missing number?

see answer
121

# Following Fun

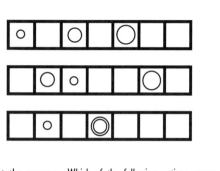

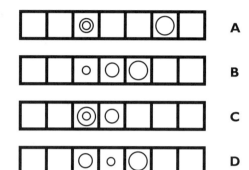

A

B

C

D

E

Look at the sequence. Which of the following options comes next?

see answer
**128**

# Triangle Teaser

Work out the three missing numbers in the third triangle.

15
75
12        19

9
45
9        18

?
15
?        ?

see answer
**122**

# Household Items

Five women each purchase a household item for use in a different room in their house.

1. Mrs Simpson does not keep her item in the bedroom or conservatory, and it is not a bookcase.
2. Amy Williams has a television; Mrs Griggs has a hi-fi.
3. The bookcase is in the study, and the television is in the bedroom.
4. Clara does not have a telephone and is not called Dingle.
5. Mrs Williams does not keep her item in the kitchen.
6. Kylie keeps hers in the conservatory.
7. Michelle has a bookcase; Mrs Dingle has a computer.
8. Clara does not keep hers in the bedroom.
9. Mrs Pringle keeps hers in the study; Roxanne keeps hers in the kitchen.

Can you work out the full name of each woman, her item and where she keeps it?

|  | FAMILY NAME | | | | | ITEM | | | | | ROOM | | | | |
|---|---|---|---|---|---|---|---|---|---|---|---|---|---|---|---|
|  | WILLIAMS | SIMPSON | PRINGLE | DINGLE | GRIGGS | TELEVISION | BOOKCASE | HI-FI | COMPUTER | TELEPHONE | LIVING ROOM | KITCHEN | CONSERVATORY | BEDROOM | STUDY |
| **FIRST NAME** KYLIE | | | | | | | | | | | | | | | |
| AMY | | | | | | | | | | | | | | | |
| CLARA | | | | | | | | | | | | | | | |
| MICHELLE | | | | | | | | | | | | | | | |
| ROXANNE | | | | | | | | | | | | | | | |
| **ROOM** LIVING ROOM | | | | | | | | | | | | | | | |
| KITCHEN | | | | | | | | | | | | | | | |
| CONSERVATORY | | | | | | | | | | | | | | | |
| BEDROOM | | | | | | | | | | | | | | | |
| STUDY | | | | | | | | | | | | | | | |
| **ITEM** TELEVISION | | | | | | | | | | | | | | | |
| BOOKCASE | | | | | | | | | | | | | | | |
| HI-FI | | | | | | | | | | | | | | | |
| COMPUTER | | | | | | | | | | | | | | | |
| TELEPHONE | | | | | | | | | | | | | | | |

| FIRST NAME | FAMILY NAME | ROOM | ITEM |
|---|---|---|---|
|  |  |  |  |
|  |  |  |  |
|  |  |  |  |
|  |  |  |  |
|  |  |  |  |

*see answer*
**118**

# Sticky Business

A stick breaks into three pieces. Without measuring the pieces or trying to construct a triangle, how can you quickly determine whether the pieces will form a triangle?

see answer
**125**

# Little and Large

Which is the odd one out?
A B C D E

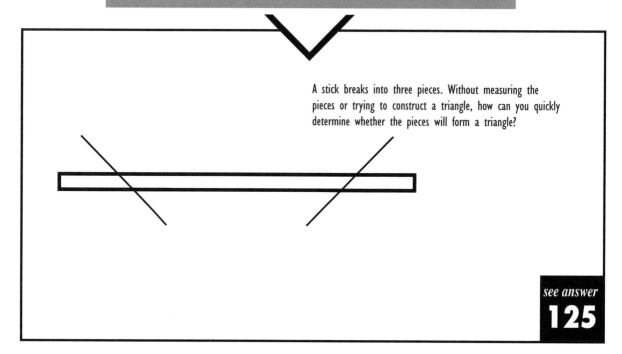

see answer
**130**

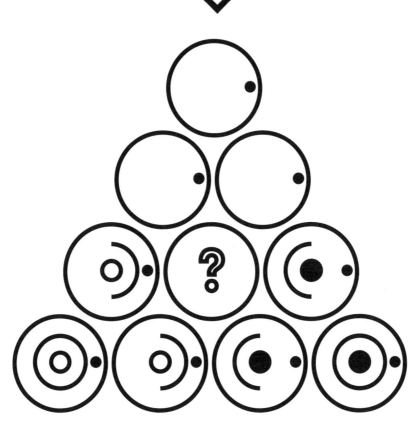

Consider the pyramid. Which of the following five options replaces the question mark?

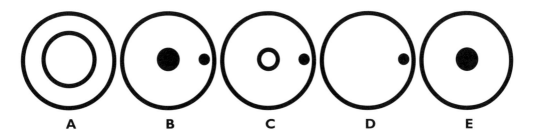

A      B      C      D      E

*see answer*
**120**

# Perpetuate the Pattern

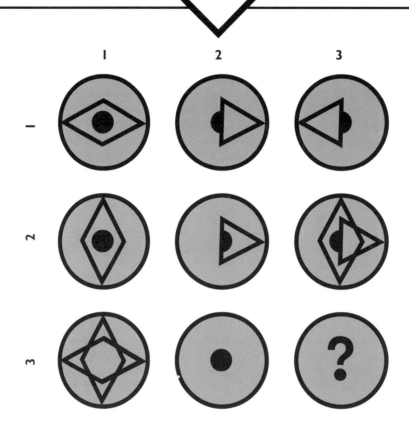

Which circle fits into the blank space to carry on the pattern?

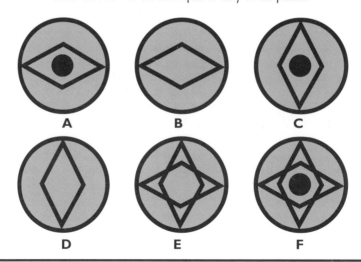

A

B

C

D

E

F

see answer
115

Each of the nine squares in the grid marked 1A to 3C should incorporate all the lines and symbols shown in the squares of the same letter and number. For example, 3C should incorporate the shapes in 3 and C.

One of the squares is incorrect. Which is it?

*see answer*
## 123

Find a logical reason for arranging these numbers into four groups of three numbers each:

**106   168   181   217   218   251   349   375   433   457   532   713**

| GROUP 1 | GROUP 2 | GROUP 3 | GROUP 4 |
|---------|---------|---------|---------|
|         |         |         |         |
|         |         |         |         |
|         |         |         |         |

*see answer*
## 126

## Answer 1
## Las Vegas

Each gambler's die was numbered as follows:

Diablo: 6 – 1 – 8 – 6 – 1 – 8
Scarface: 7 – 5 – 3 – 7 – 5 – 3
Lucky: 2 – 9 – 4 – 2 – 9 – 4

In a long run:

Diablo would win against Scarface 10 times in 18;
Scarface would win against Lucky 10 times in 18;
Lucky would win against Diablo 10 times in 18.

Diablo v Scarface: 6–7; 1–7; 8–7 win; 6–5 win; 1–5; 8–5 win; 6–3 win; 1–3; 8–3 win, which, when repeated, gives 10 wins and 8 losses for Diablo.
Scarface v Lucky: 7–2 win; 5–2 win; 3–2 win; 7–9; 5–9; 3–9; 7–4 win; 5–4 win; 3–4, which, when repeated, gives 10 wins and 8 losses for Scarface.
Lucky v Diablo: 2–6; 9 – 6 win; 4–6; 2–1 win; 9–1 win; 4–1 win; 2–8; 9–8 win; 4–8, which, when repeated, gives 10 wins and 8 losses to Lucky.

## Answer 2
## Three Circles

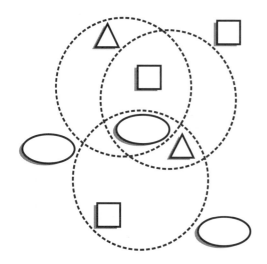

## Answer 3
## Town Clock

I forgot to mention that my clock was digital. One line was not functioning on the eight lines that make up each digit.

THIS LINE NOT FUNCTIONING

| | TIME SHOWED | SHOULD HAVE SHOWN |
|---|---|---|
| 8.55 | 5 | 5 |
| 8.56 | 6 | 6 |
| 8.58 | 6 ◀ MISSING | 8 |
| 8.59 | 5 ◀ MISSING | 9 |
| 9.00 | ⌐ ◀ MISSING | 0 |

## Answer 4
## Booth Bonanza

If number eight did not require repairing the supervisor should have said that five out of the first seven needed repairing.

## Answer 5
## Tricky Triangles

E. There are four triangles constantly moving clockwise around the arms and visiting points in sequence.

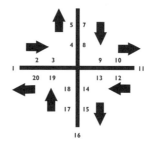

## Answer 6
## Murder in Mind
Alf Muggins. If it was Jack Vicious, the statements of Alf Muggins and Jim Pouncer would be true. If it was Sid Shifty, the statements of Jack Vicious, Alf Muggins and Jim Pouncer would be true. If it was Jim Pouncer, the statements of Sid Shifty and Alf Muggins would be true. Therefore it is Alf Muggins, and only the statement of Jim Pouncer is true.

## Answer 7
## Fancy Figures
2, 5, 7, 10, 12, 15, 17.

## Answer 8
## Hexagon Harmony
A. There are six triangles, each with their base on one of the sides of the hexagon. Each triangle increases in height by a quarter of the width of the hexagon at each stage. So, showing one triangle only:

## Answer 9
## Manor House

| Name | Occupation | Pastime | Rest day |
|------|-----------|---------|----------|
| Smith | Butler | Squash | Friday |
| Jones | Gardener | Golf | Tuesday |
| Wood | Chauffeur | Fishing | Wednesday |
| Clark | Janitor | Chess | Thursday |
| James | Cook | Bridge | Monday |

## Answer 10
## Converging Circles
C.

## Answer 11
## Sears Tower
450m.

## Answer 12
## Land of Zoz

| What they say they are | Numbers in the group | What they actually are | What they become |
|------------------------|---------------------|------------------------|------------------|
| Fibkins | 30 | 30 Switchkins | 30 Fibkins |
| Switchkins | 15:15 | 15 Fibkins | 15 Fibkins |
| | | 15 Switchkins | 15 Switchkins |
| Truthkins | 10:10:10 | 10 Truthkins | 10 Truthkins |
| | | 10 Fibkins | 10 Fibkins |
| | | 10 Switchkins | 10 Truthkins |

Therefore 55 Fibkins live in pentagonal houses that night.

Truthkins only tell the truth, and must be included in the group of 3 x 10. This group therefore made the statement "We are all Truthkins." The Switchkins must be the group of 30 x 1, which lied "We are all Fibkins", thus becoming Fibkins – it could not be Fibkins making this statement because they always lie. This means that the other statement was made by 15 x 2 of Switchkins and Fibkins. Therefore, there are 55 Fibkins (10 who lie about being Truthkins, the 15 who claimed to be Switchkins and the 30 former Switchkins who became Fibkins by claiming they were Fibkins).

## Answer 13
## Making Eyes
D.
There is a sequence occurring from the right eye to the left eye (as we look at them). Look at stages one and two. The contents of the eyes in stage one have merged to form the left eye of stage two and a new symbol has been introduced in the right eye of stage two. Now look at stages two and three. The contents of the left eye in stage two has moved away and does not appear in stage three. The symbol from the right eye in stage two has moved to fill the left eye of stage three and a new symbol has been introduced in the right eye of stage three. This pattern of change is then continued, so that the left eye of stage four contains a merging of both eyes in stage three.

## Answer 14
### Shooting Range
Colonel Present scored 200 ( 60, 60, 40, 40)
Major Aim scored 240 (60, 60, 60, 60)
General Fire scored 180 (60, 40, 40, 40)

The incorrect statements made by each marksman are as follows:
Colonel Present, statement no. 1;
Major Aim, statement no. 3;
General Fire, statement no. 3.

## Answer 15
### Counterfeit Coins
Only one weighing operation is necessary. You take one coin from bag one, two coins from bag two and three coins from bag three and weigh all six coins together. If they weigh 305g the first bag contains the counterfeit coins; if they weigh 310g the second bag does, and if they weigh 315g the third bag does.

## Answer 16
### Sitting Pretty

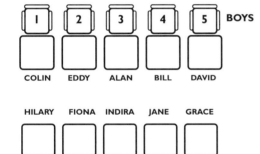

## Answer 17
### Figure of Fun
C.
The number of right angles in each figure increases by one each time.

## Answer 18
### House Hunting
We have to find a unique number which will answer three of the questions.

1. Is it under 41?

No 41–82

Yes 1-40

2. Is it divisible by 4?

| Yes | 44 | 48 | 52 | 56 | 60 | 64 | 68 | 72 | 76 | 80 |
|---|---|---|---|---|---|---|---|---|---|---|
| No | 41 | 42 | 43 | 45 | 46 | 47 | 49 | 50 | 51 | 53 |
|  | 54 | 55 | 57 | 58 | 59 | 61 | 62 | 63 | 65 | 66 |
|  | 67 | 69 | 70 | 71 | 73 | 74 | 75 | 77 | 78 | 79 |
|  | 81 | 82 |  |  |  |  |  |  |  |  |
| Yes | 4 | 8 | 12 | 16 | 20 | 24 | 28 | 32 | 36 | 40 |
| No | 1 | 2 | 3 | 5 | 6 | 7 | 9 | 10 | 11 | 13 |
|  | 14 | 15 | 17 | 18 | 19 | 21 | 22 | 23 | 25 | 26 |
|  | 27 | 29 | 30 | 31 | 33 | 34 | 35 | 37 | 38 | 39 |

3. Is is a square number?

Yes  64 unique

| No | 44 | 48 | 52 | 56 | 60 | 68 | 72 | 76 | 80 |  |  |
|---|---|---|---|---|---|---|---|---|---|---|---|
| Yes | 49 | 81 |  |  |  |  |  |  |  |  |  |
| No | 41 | 42 | 43 | 45 | 46 | 47 | 50 | 51 | 53 |  |  |
|  | 54 | 55 | 57 | 58 | 59 | 61 | 62 | 63 | 65 | 66 |  |
|  | 67 | 69 | 70 | 71 | 73 | 74 | 75 | 77 | 78 | 79 | 82 |
| Yes | 4 | 16 | 36 |  |  |  |  |  |  |  |  |
| No | 8 | 12 | 20 | 24 | 28 | 32 | 40 |  |  |  |  |
| Yes | 1 | 9 | 25 |  |  |  |  |  |  |  |  |
| No | 2 | 3 | 5 | 6 | 7 | 10 | 11 | 13 |  |  |  |
|  | 14 | 15 | 17 | 18 | 19 | 21 | 22 | 23 |  |  |  |
|  | 26 | 27 | 29 | 30 | 31 | 33 | 34 | 35 | 37 | 38 | 39 |

By answering no to the first question, yes to the second and yes to the third you arrive at a unique number – 64, which is therefore the number of Archibald's house.

## Answer 19
### Dynamic Dog
22½ miles. Work out how long it takes Russell Carter to walk home. Spot has been running all this time at his given constant speed so it is simple to work out how many miles Spot has covered during this period.

Russell walks for 10 miles at 4mph, taking 2½ hours. Spot is running for 2½ hours too, at 9mph, which means he covers 22½ miles.

## Answer 20
### Roving Robot
There was a stationary car parked 5 m to the robot's right. The program should have said "moving land vehicle".

## Answer 21
### Dazzling Diamond

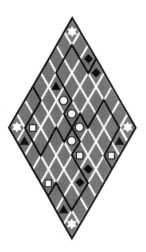

## Answer 22
### Japan Hotel
"Push" on one side; "Pull" on the other side.

## Answer 23
### Salary Increase
At first glance the best option seems to be the first. However, the second option works out best.

| First option | ($500 increase after each 12 months) |
|---|---|
| First year | $10,000 + $10,000 = $20,000 |
| Second year | $10,250 + $10,250 = $20,500 |
| Third year | $10,500 + $10,500 = $21,000 |
| Fourth year | $10,750 + $10,750 = $21,500 |

| Second option | ($125 increase after each 6 months) |
|---|---|
| First year | $10,000 + $10,125 = $20,125 |
| Second year | $10,250 + $10,375 = $20,625 |
| Third year | $10,500 + $10,625 = $21,125 |
| Fourth year | $10,750 + $10,875 = $21,625 |

## Answer 24
### Funny Fingers
Let us assume that 240 fingers could be 20 aliens with 12 fingers each or 12 aliens with 20 fingers each, etc. This does not provide a unique answer so eliminates all numbers that can be factorized.

Now consider prime numbers: there could be one alien with 229 fingers (not allowed, according to sentence one); 229 aliens with one finger (not according to sentence two). Again, this does not provide a unique answer so eliminates all prime numbers except those squared. There is only one of these between 200 and 300 and that is 289 ($17^2$). So in the room are 17 aliens each with 17 fingers.

## Answer 25
### Pick a Pattern
C. The middle pattern is removed and encases the outer pattern of the original figure.

## Answer 26
### Barrels of Fun
The 40-gallon barrel contains beer. The first customer purchases the 30-gallon and 36-gallon barrels, giving 66 gallon of wine. The second customer purchases 132 gallons of wine – the 32-gallon, 38-gallon and 62-gallon barrels. The 40-gallon barrel has not been purchased by either customer and therefore contains the beer.

## Answer 27
### Mouse Moves
As there are 216 chambers – an even number – there is no central chamber. The task is therefore impossible.

## Answer 28
### Treasured Trees

| Tree | Elm | Ash | Beech | Lime | Poplar |
|---|---|---|---|---|---|
| Person | Bill | Jim | Tony | Sylvester | Desmond |
| Club | Squash | Golf | Tennis | Bowling | Soccer |
| Bird | Owl | Blackbird | Crow | Robin | Starling |
| Year | 1970 | 1971 | 1972 | 1973 | 1974 |

## Answer 29
### Intergalactic Ingenuity
The wife of M9 is F10. The male speaker on nuclear fission is M3.

| Male | M1 | M3 | M5 | M9 | M7 |
|---|---|---|---|---|---|
| Female | F8 | F6 | F4 | F10 | F2 |
| Vehicle | Warp distorter | Galaxy freighter | Space oscillator | Nebula accelerator | Astro carrier |
| Speech | Time travel | Nuclear fission | Astral transporting | Mind reading | Anti-gravity |
| Feature | 12 fingers | 3 eyes | 3 legs | 4 arms | Webbed feet |

## Answer 30
### Handkerchief Challenge
Charlie puts the handkerchief under a door and stands on the corner at the other side.

## Answer 31
### Algarve Rendezvous

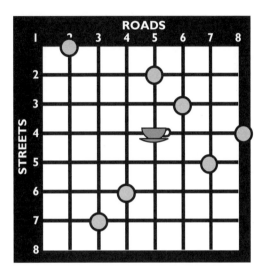

On the corner of road 5, street 4. Draw a line down the person who is in the middle on the roads axis. Then draw a line across the person who is in the middle of the streets axis.

## Answer 32
### Searching Segments

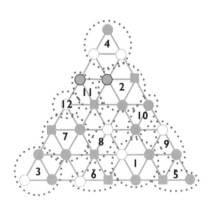

## Answer 33
## Take a Tile
**B.**
Looking both across and down, any lines common to the first two tiles disappear in the third tile.

## Answer 34
## Prisoners' Porridge

2519 prisoners.

2519 divided by 3 = 839 tables with 2 over
2519 divided by 5 = 503 tables with 4 over
2519 divided by 7 = 359 with 6 over
2519 divided by 9 = 279 with 8 over
2519 divided by 11 = 229 exactly.

## Answer 35
## Creative Circles
A. Looking both across and down, the contents of the third square are formed by merging the contents of the two previous squares as follows:

one white or black circle remains;
two black circles become white;
two white circles become black.

## Answer 36
## Think of a Number
Anastasia tells a lie when she says that the number is below 500. The only square and cube between 99 and 999 whose first and last digit is 5, 7 or 9 is 729.

## Answer 37
## Dog Delight
The dalmations are called Andy (owned by Bill) and Donald (owned by Colin).

## Answer 38
## Club Conundrum
2. If all 49 women wore glasses then 21 men wore glasses too. If 11 of these men were under 20 years of age, only 10 men older than 20 years of age wore glasses. Then 10 – 8 = 2 men is the minimum number.

## Answer 39
## Three Squares

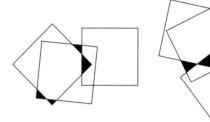

**D.**
The three squares form four triangles.

## Answer 40
## Roulette Riddle
15.

## Answer 41
## Tree Teaser
Spencer prunes six more trees than Don.

## Answer 42
## Girl Talk

| The three ages, when multiplied, must be one of the following combinations: | When added, they equal |
|---|---|
| 72 x 1 x 1 | 74 |
| 36 x 2 x 1 | 39 |
| 18 x 4 x 1 | 23 |
| 9 x 4 x 2 | 15 |
| 9 x 8 x 1 | 18 |
| 6 x 6 x 2 | 14 |
| 8 x 3 x 3 | 14 |
| 12 x 6 x 1 | 19 |
| 12 x 3 x 2 | 17 |
| 18 x 2 x 2 | 22 |
| 6 x 3 x 4 | 13 |
| 3 x 24 x 1 | 28 |

The census-taker should have known the number of the house, as he could see it, but he did not know their ages, therefore the house must be number 14. He needed more information to decide whether their ages were 6, 6, 2 or 8, 3, 3. When the woman says "eldest" daughter, he knows they were 8, 3, 3.

## Answer 43
## Lucid Lines
E.
The two figures merge into one by superimposing one onto the other, except that when two lines appear in the same position they disappear.

## Answer 44
## Number Placement

| 2 | 14 | 10 | 7 |
|---|----|----|---|
| 9 | 6 | 1 | 4 |
| 16 | 3 | 13 | 11 |
| 12 | 8 | 5 | 15 |

So that:
1. No two consecutive numbers appear in any horizontal, vertical or diagonal line;
2. No two consecutive numbers appear in adjacent squares.
Note also the positions of 1 and 2 could be swapped.

## Answer 45
## Play Watching

 MR

 MRS

## Answer 46
## Suspicious Circles
E. A is a mirror image of C; B is a mirror image of D.

## Answer 47
## Study Time
Anne studies algebra, history, French and Japanese.
Bess studies physics, English, History and Japanese.
Candice studies algebra, physics, English and French.

## Answer 48
## Tricky Triangles

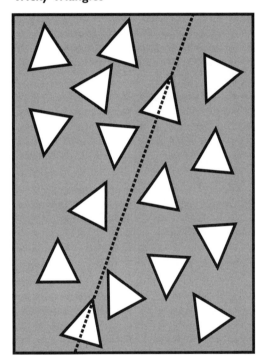

Two segments: all the triangles do not need to be the same size.

## Answer 49
## Picking Professions
Mr Carter is the drover.

### Answer 50
### Great Golfers
Edward Peters; Robert Edwards; Peter Roberts. Mr Peters must be Edward because the man who spoke last is Robert and he is not Mr Peters.

### Answer 51
### Lost Time

$11 + 12 + 1 + 2 = 26$
$10 + 3 + 9 + 4 = 26$
$5 + 6 + 7 + 8 = 26.$

### Answer 52
### Pleased Pupils

| Name | Class | Subject | Sport |
|---|---|---|---|
| Alice | 6 | Algebra | Squash |
| Betty | 2 | Biology | Running |
| Clara | 4 | History | Swimming |
| Doris | 3 | Geography | Tennis |
| Elizabeth | 5 | Chemistry | Basketball |

### Answer 53
### Hello Sailor

| Name | Rank | Ship | Location |
|---|---|---|---|
| Perkins | Steward | Aircraft carrier | Malta |
| Ward | Seaman | Cruiser | Portsmouth |
| Manning | Commander | Submarine | Falklands |
| Dewhurst | Purser | Frigate | Gibraltar |
| Brand | Captain | Warship | Crete |

### Answer 54
### Gone Fishing

North ← Pier → South

| Name | Joe | Fred | Dick | Henry | Malcolm |
|---|---|---|---|---|---|
| Occupation | Banker | Electrician | Professor | Plumber | Salesman |
| Town | L.A. | Orlando | Tucson | New York | St Louis |
| Bait | Worms | Bread | Maggots | Shrimps | Meal |
| Catch | 6 | 15 | 10 | 9 | 1 |

### Answer 55
### Winning Wager
It is possible but there has to be a compensating factor. Jim has $8 at the start so Bill can win only $8 even if he wins all 10 frames. Jim, however, can win a large sum if he wins every frame: $8, $12, $18, $27, etc. Therefore, to compensate, Bill can win a small amount even if he wins fewer frames than Jim. If makes no difference in the winning order of frames.

| Frame | Jim | Jim has $8 |
|---|---|---|
| 1 | win | $12.00 |
| 2 | lose | $6.00 |
| 3 | lose | $3.00 |
| 4 | win | $4.50 |
| 5 | win | $6.75 |
| 6 | lose | $3.38 |
| 7 | win | $5.07 |
| 8 | win | $7.60 |
| 9 | win | $11.40 |
| 10 | lose | $5.70 = loses $2.30 from starting $8.00 |

**Answer 56**
**Odd One Out**
E. In all the others, if the line dividing the square is a mirror the correct mirror image has been shown.

**Answer 57**
**Spot the Shape**
1A.

**Answer 58**
**Strange Series**

If you don't believe this, hold the book up to a mirror. You will see that with the inclusion of the above, the numbers 1, 2, 3, 4, 5 appear in sequence.

**Answer 59**
**Changing Trains**
25 minutes. As the man leaves home according to his normal schedule it is earlier than 6.30 pm when he picks up his wife. As the total journey saves 10 minutes, that must be the same time it takes the man from the point he picks up his wife to the station and back to the same point. Assuming that it takes an equal five minutes each way he has therefore picked up his wife five minutes before he would normally, which means 6.20pm. So his wife must have walked from 6.00pm to 6.25pm, that is for 25 minutes.

**Answer 60**
**Fathers and Daughters**

| Father | Daughter | Father's age | Daughter's age |
|--------|----------|--------------|----------------|
| John | Alison | 52 | 20 |
| Kevin | Diana | 53 | 19 |
| Len | Betty | 50 | 21 |
| Malcolm | Eve | 54 | 18 |
| Nick | Carol | 51 | 17 |

**Answer 61**
**Sums about Sons**
Graham is 9 years old and Frederick is 27. Thus, 27 squared is the same as 9 cubed = 729. There are 18 steps, 36 palisades and 243 bricks, which, when added together, gives the door number of 297.

**Answer 62**
**Cube Diagonals**
60°. If a third face diagonal, BC, is drawn this completes an equilateral triangle. All its sides are equal because they are cube diagonals. Being equilateral, all its angles are 60°.

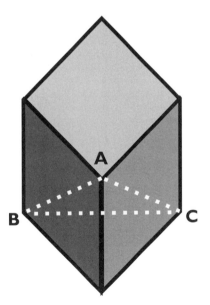

## Answer 63
### Best Beer
18 days. If it takes a man 27 days to drink one barrel, he drinks 0.037 of a barrel each day. Similarly, a woman drinks 0.0185 of a barrel each day. Added together, a day's combined drinking consumes 0.0555 of a barrel. In this case, to drink the whole barrel takes 18.018 days.

## Answer 64
### Circles and Triangles
C. Each horizontal line and vertical column contains the wavy shape shown once vertical and once black. Similarly, each line and column shows the triangle three times: once pointing left, once right, once down and once black and twice right.

## Answer 65
### Gun Running
The total number of dollars that they receive for their cattle must be a square number. They buy an odd number of sheep at $10 each, so the tens figure in the total square number must be an odd number. The only square numbers with an odd "tens" figure have "6" as their "units" figure.

The number 256 is one such number, equalling the price of 16 steer at $16 a head as well as 25 sheep at $10 a head with $6 for the goat. Bully Bill evens up the takings by giving Dynamo Dan a goat (worth $6) and the Colt .45 to equal his own share of the sheep ($10) minus the Colt .45. Therefore the gun is worth $2.

## Answer 66
### Missing Numbers
The grid should contain 1x1, 2x2, 3x3, 4x4, 5x5, 6x6, 7x7, 8x8. Therefore, the missing numbers are 2, 7, 7, 8. All the numbers are placed so that two identical numbers are never adjacent.

## Answer 67
### Rifle Range
Major Mustard. Tabulate the results so that each set equals 71. There are only three possible ways to do this: 25, 20, 20, 3,2,1; 25, 20, 10, 10, 5, 1; and 50, 10, 5, 3, 2, 1. The first set is Colonel Ketchup's (since 22 cannot be scored in two shots in the other sets); the third set is Major Mustard's (as we know that he scores 3). So Major Mustard hit the bull's eye.

## Answer 68
### Round the Hexagons
C.

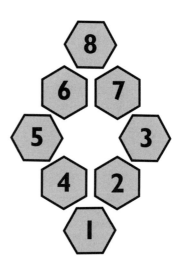

The third hexagon is formed by merging hexagons 1 and 2. The fifth hexagon is formed by merging hexagons 4 and 1. In this way, the hexagons build up the shape along vertical lines going from the bottom hexagon upwards. Continuing this trend, the top hexagon is formed by a merging of hexagons 3, 5, 6, and 7: the two straight lines moving upwards to the top hexagon.

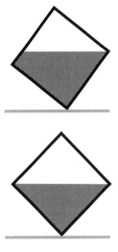

**Answer 69**
**Skyscraper Sizzler**
She lives on the 27th floor. The elevator came down from the 36th to the 28th floor – 9 floors; or it came up from the first to the 27th floor – 27 floors. Therefore there is a 3:1 chance of it going up than down.

**Answer 70**
**Black and White Balls**
Three chances in four. Look at the possible combinations of drawing the balls. There are black–black; white–black; black–white; and white–white. The only one of the four possible combinations in which it does not occur is the fourth one. The chances of drawing at least one black ball are, therefore, three in four.

**Answer 71**
**Carrier Pigeons**
No. The pigeons remain at 200lbs even whilst flying. Those flying up would reduce the weight, but those flying down would increase the weight, so balancing the total weight.

**Answer 72**
**Bartender's Beer**
The first man places a $1 bill on the counter. The second man puts down three quarters, two dimes and a nickel – amounting to one dollar. Had he wanted a 90-cent beer he had the change to offer the exact amount.

**Answer 73**
**Logical Clocks**
A. At each stage the big hand moves anti-clockwise first by 10 minutes, then 20 and, finally, by 30 minutes (option A). At each stage the small hand moves clockwise first by one hour, then two hours and, finally, three hours (option A).

**Answer 74**
**Broadway, NY**
The fact that the man does not see a door (as in the illustration) indicates that the door must be on the other side – the kerb side. As this is New York, the bus is therefore moving to A.

**Answer 75**
**Water Divining**

By lifting the water tank onto its near-side edge.
If you cannot see the far edge then the tank is more than half full.
If you can just see the far edge then the tank is exactly half full.
If you can see below the far edge then the tank is less than half full.

### Answer 76
### Missing Links

1. In the first bar, 7 x 4 x 8 x 8 x 2 = 3584.
Similarly, 3 x 5 x 8 x 4 = 480.
Following the same formula, the missing number in the second bar is 2268 and, in the third, 2688 and 768.

2. In the first bar 58 x 2 = 116.
In the same vein, 16 x 1 = 16. Using the same formula, the missing number in the second bar is 657 and, in the third, 162 and 72.

### Answer 77
### City Slicker

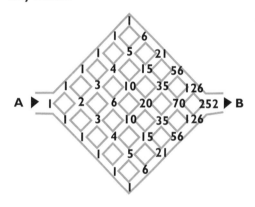

252. Each number represents the number of possible routes to each intersection.

### Answer 78
### Star Gazing

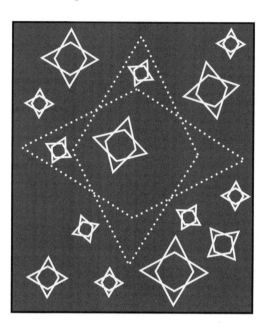

### Answer 79
### Pyramid Puzzle
D.
Each pair of circles produces the circle above by carrying forward only those elements that are different. Similar elements disappear.

## Answer 80
## Round and Round the Garden

722m.
The path takes up the area of the garden, or 1,444m.
It is 2m wide, so it's length is 722m.

## Answer 81
## Door Number Puzzle
No in both cases, except if the 6 is turned upside down into a 9. If the numbers were 1, 4, 7 and 9, it could never be divisible by 9 but always by 3.

## Answer 82
## Dice Dilemma
No. 3.

## Answer 83
## Triangles and Trapeziums
A.
The figures change places so that the one in front goes to the back and vice versa.

## Answer 84
## Ski-lift
90.
You can buy nine tickets from each of the 10 stations: 9 x 10 = 90.

## Answer 85
## Five Pilots

| Name | Airport | Destination |
|------|---------|-------------|
| Mike | Heathrow | JFK |
| Nick | Gatwick | Vancouver |
| Paul | Cardiff | Berlin |
| Robin | Manchester | Roma |
| Tony | Stansted | Nice |

## Answer 86
## Making Moves
J is option 4; N is option 6. The black segments move from top to bottom and right to left in sequence, then rise in the same way. However, when an arrangement has occurred previously it is omitted from the sequence.

## Answer 87
## Dinner Party Placements
Mr and Mrs Chester.

## Answer 88
## Careful Calculation
8679.
Turn the page upside-down and add up the two numbers.

## Answer 89
## Pyramidal Logic.
E. Each symbol is linked to the two below it. No symbol never appears above an identical one. The symbols are produced as follows:

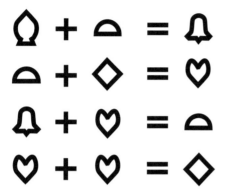

so that + ⬯ must equal something completely different to anything above. Of the options shown, this can only be ♧.

## Answer 90
## Figure Columns
D. The smallest number is dropped each time and the remaining numbers appear in reverse order.

## Answer 91
## Sun Shine

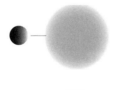

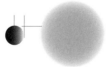

Any valley on or near the equator, owing to the revolution of the Earth.

## Answer 92
## Shady Squares
C.
The square turns 90° clockwise at each stage. Similarly, the shading also moves one segment clockwise at each stage.

## Answer 93
## Generation Gap
I am 40 and my daughter is 10.

## Answer 94
## Work it Out
24. In the first circle, 56 + 79 divided by 5 = 27. The same formula applies to circles two and three.

## Answer 95
## Chess Strategy
Strong, weak, strong. Assuming he is likely to beat the weak player, playing this way gives him two chances to beat the strong one.

## Answer 96
## Lonely Loser
E.
The others all have rotated symmetry. In other words, if they were rotated through 180° they would appear exactly the same.

## Answer 97
## Scratch Card
The number of squares on the card is immaterial. The odds are always 2:1 against.

## Answer 98
## Frogs and Flies
29.

## Answer 99
### Lateral Logic
B.

There are three sizes of rectangle. In the next three stages A moves from left to right one stage at a time. Then it is the turn of B to do the same.

## Answer 100
### Missing Number
12.

The third number, 27, is obtained by adding the digits of the two preceding numbers – 7 + 2 + 9 + 9. This formula applies throughout the puzzle.

## Answer 101
### Pentagon Figures
4. In the first pentagon 5 x 5 x 125 = 3125 or $5^5$. In the second pentagon, 3 x 9 x 9 = 243 or $3^5$. In the same way, 16 x 8 x 8 = 1024 or $4^5$.

## Answer 102
### Counting Creatures
44.

28 huskies with four legs each plus 44 penguins with two each, making 200 in all.

## Answer 103
### Eighteen Trees

Solution 1

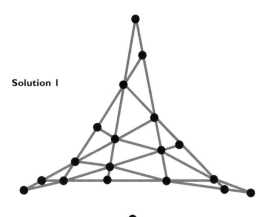

Solution 2

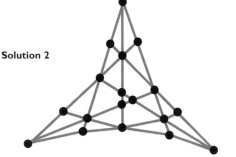

Both solutions produce nine rows of five trees per row.

## Answer 104
### Fairground Fiesta

| Name | Age | Ride | Food |
|------|-----|------|------|
| Sam | 14 | Dodgems | Hot dog |
| Joe | 11 | Big dipper | Fries |
| Don | 12 | Whirligig | Candy floss |
| Len | 15 | Crocodile | Gum |
| Ron | 13 | Mountain | Ice cream |

### Answer 105
### Five Circles

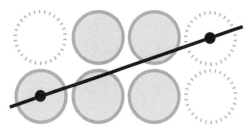

### Answer 106
### Line Analogy
B. The figures are flipped vertically.

### Answer 107
### Island Access

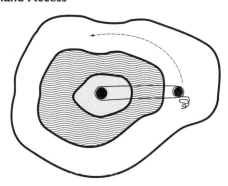

He ties the rope to the tree, then walks around the lake carrying the rope. As he reaches half-way, the rope wraps itself around the tree on the island. He then ties the rope to the tree on the mainland and hauls himself across to the island.

### Answer 108
### Number Crunching
33.
Multiply diagonally opposite squares and subtract the smaller product from the larger:
(13 x 5) - (8 x 4) = 33.

### Answer 109
### Square Sort
2C.

### Answer 110
### Round in Circles
C.
The striped and black segments are moving in the following sequences: the striped segments move two anti-clockwise then one clockwise in turn, and continue in this way. The black segments move two clockwise then one anti-clockwise in turn, and continue in this way too.

### Answer 111
### Logic Circles
D.
The large circle moves 180°; the small white circle moves 180°; the black circle moves 90°; and the black dot moves 180°.

### Answer 112
### Easy Equation
$7^2$ = 49. The 6 has been turned over to convert it into a 9 and the 2 becomes a square.

### Answer 113
### Suspicious Shape
B.
A and F are the same, as are C and D, and E and G.

### Answer 114
### Bird Fanciers

| Name | Country | Birds | Collective Noun |
|---|---|---|---|
| Albert | Belgium | Owls | Parliament |
| Roger | France | Crows | Murder |
| Harold | Germany | Ravens | Unkindness |
| Cameron | Scotland | Plovers | Wing |
| Edward | England | Starlings | Murmuration |

## Answer 115
### Perpetuate the Pattern
F.
Column 1 is added to column 2 to make column 3. Similarly, line 1 is added to line 2 to make line 3. In both cases, repeated symbols disappear.

## Answer 116
### Hexagonal Pyramid
E.
The contents of each hexagon are determined by merging the contents of the two hexagons immediately below, except that two identical lines disappear.

## Answer 117
### Sequence Search
D.

 moves 135° clockwise

 moves 45° clockwise

 moves 90° clockwise

 moves 180°

## Answer 118
### Household Items

| First name | Family name | Room | Appliance |
|---|---|---|---|
| Kylie | Dingle | Conservatory | Computer |
| Amy | Williams | Bedroom | Television |
| Clara | Griggs | Living Room | Hi-fi |
| Roxanne | Simpson | Kitchen | Telephone |
| Michelle | Pringle | Study | Bookcase |

## Answer 119
### Trying Trominoes
B.
There are four different symbols grouped ABC, ABD, BCD and, in option B, ACD.

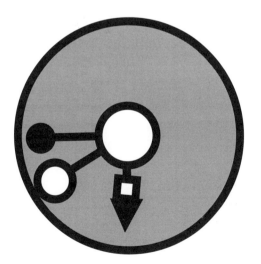

A       B       C       D

**Answer 120**
**Pyramid Plot**
D.
Each pair of circles produces the circle above by carrying on elements that they have in common. Different elements disappear.

**Answer 121**
**Round the Circle**
20.
Start at 10 and jump to alternate segments, adding 1, then 2, then 3 and so on.

**Answer 122**
**Triangle Teaser**

Divide the central number by 5 to give the top number. Add the digits of the central number to give the bottom left number. Reverse the digits of the central number and divide by three to give the bottom right number.

**Answer 123**
**Gritty Grid**
3A.

**Answer 125**
**Sticky Business**
If the shorter pieces, placed end to end, are longer than the largest piece, then they will form a triangle.

**Answer 126**
**Fancy Figures**
Arrange them into groups of three, each totalling 1000.

457 + 168 + 375 = 1000

532 + 217 + 251 = 1000

349 + 218 + 433 = 1000

713 + 106 + 181 = 1000

**Answer 127**
**Unwanted Guest**
B.
It is a straight-sided figure within a curved figure.
The rest are curved figures within a straight
sided one.

**Answer 128**
**Following Fun**
D.
The small circle moves two on and one back. The
middle-size circle moves one back and two on. The
large circle moves one on and two back.

**Answer 129**
**Symbol Search**
B.

**Answer 130**
**Little and Large**
C. B and D, and A and E are the same, with large
and small circles reversed.

# Suit Trick

Consider the arrangement of suits of cards above.

Taking into account the eight other arrangements above, which of the following arrangements should replace the question marks?

see answer

1

# Spelling Bee

A school needed to pick a team of 11 children for a spelling competition. They had to select 5 boys and 6 girls from a total of 10 boys and 11 girls.

How many different teams could they have chosen?

see answer
9

# Eleven-Tree Shuffle

A gardener has 11 trees that he wants to plant in straight rows of three trees per row. He also wants to plant the 11 trees in such an arrangement that he will obtain the maximum possible number of rows at three trees per row. How does he do it? The example shows the maximum six rows possible with just seven trees.

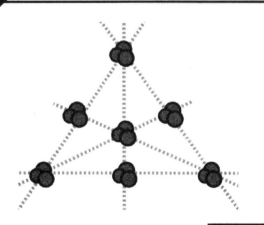

see answer
13

Which is the odd one out?

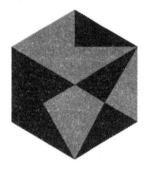

**A**

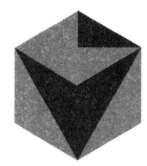

**B**

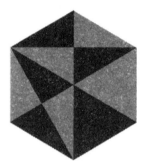

**C**

**D**

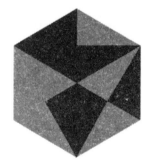

**E**

**F**

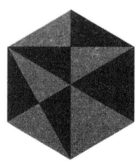

**G**

*see answer*

**10**

# Circular Question

What number should replace the question mark?

see answer
**14**

# Shape Up

Which arrangement continues this sequence?

see answer
**4**

A     B     C     D     E

# Circular Connection

If 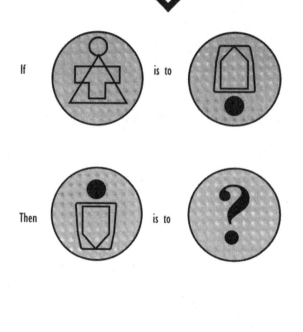 is to

Then is to ?

A

B

C

D

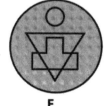

E

see answer
**21**

## Likely Coins

Someone tosses five coins in the air at the same time and you are betting on the outcome. What are the chances:
i) that at least three of the coins will finish up either all heads or all tails?
ii) that at least three of the coins will finish up all heads?
iii) that at least four of the coins will finish up all heads?

*see answer*
**3**

## Keyboard Teaser

My word processor keyboard had a sticky key. I was not using the machine but at various times the six people in my family were. I asked each person to name the key so that I could have it repaired. They answered:

A said, "It was *, $, or &"
B said, "It was —, a, or !"
C said, "It was K, ?, or %"
D said, "It was }, a, or &"
E said, "It was K, $, or !"
F said, "It was %, ?, or K"

But half of them had lied. Which key was it?

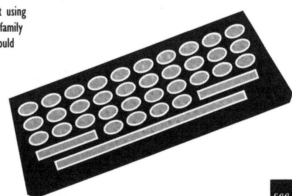

*see answer*
**8**

# Truth Chase

My friend ran in a steeplechase of 16 runners. I missed the race and asked six spectators to tell me in which position my friend had finished. These were the replies:

A said, "It was a single digit"
B said, "It was a double digit"
C said, "It had at least one number 1 in it"
D said, "It had a 5 in it"
E said, "It was between 5 and 11"
F said, "It was an even number"

Only four people had told the truth. What position had my friend finished?

see answer
5

# Democratic Digits

At a general election, a total of 96,284 votes were cast for the four candidates. The winner exceeded her opponents by 5,392, 7,845 and 10,219 votes respectively.

How many votes were received by each candidate?

see answer
12

# Who's Dancing?

Five men went to a fancy dress ball. Each went as a different character, had a different costume, and danced a different dance.

Simon wore a Fedora, Henry danced the Bossa Nova, Dr Jekyll Jitterbugged, Robert went as Napoleon, Morris went as Shakespeare, Peter wore a Homburg, Frankenstein danced the Palais Glide, Peter danced the Palais Glide. The man in jackboots danced the Charleston. The man who danced the Barn Dance wore a skull cap. Dracula wore leggings, Robert wore jackboots.

Can you match the men to their character, costume and dance?

|  |  | FANCY DRESS | | | | | CLOTHES | | | | | DANCE | | | | |
|---|---|---|---|---|---|---|---|---|---|---|---|---|---|---|---|---|
|  |  | DRACULA | NAPOLEON | DR JEKYLL | FRANKENSTEIN | SHAKESPEARE | LEGGINGS | JACK BOOTS | FEDORA | HOMBURG | SKULL-CAP | BOSSA-NOVA | CHARLESTON | JITTERBUG | PALAIS GLIDE | BARN DANCE |
| NAME | SIMON | | | | | | | | | | | | | | | |
|  | MORRIS | | | | | | | | | | | | | | | |
|  | HENRY | | | | | | | | | | | | | | | |
|  | PETER | | | | | | | | | | | | | | | |
|  | ROBERT | | | | | | | | | | | | | | | |
| DANCE | BARN DANCE | | | | | | | | | | | | | | | |
|  | CHARLESTON | | | | | | | | | | | | | | | |
|  | PALAIS GLIDE | | | | | | | | | | | | | | | |
|  | BOSSA-NOVA | | | | | | | | | | | | | | | |
|  | JITTER BUG | | | | | | | | | | | | | | | |
| CLOTHES | HOMBURG | | | | | | | | | | | | | | | |
|  | SKULL-CAP | | | | | | | | | | | | | | | |
|  | FEDORA | | | | | | | | | | | | | | | |
|  | JACK BOOTS | | | | | | | | | | | | | | | |
|  | LEGGINGS | | | | | | | | | | | | | | | |

| NAME | FANCY DRESS | CLOTHES | DANCE |
|---|---|---|---|
|  |  |  |  |
|  |  |  |  |
|  |  |  |  |
|  |  |  |  |
|  |  |  |  |

*see answer*
**2**

# Spot the Number

| 21 | 8 | 9 |
| 32 | 3 | 9 |
| 15 | 3 | ? |

What number should replace the question mark?

Clue: Look for two different divisors.

see answer
19

# Red Car Blues

In one week a car showroom was selling only blue and red cars. Throughout the week the same number of cars were for sale — when one was sold it was immediately replaced by another. A red car was sold first and was replaced by a blue one, after which there were the same number of red and blue cars for sale. Later a blue car was sold and replaced by a red one, giving the same amount of red and blue cars as originally. Then another blue car was sold and replaced by a red one. There were then twice as many red cars as blue ones.

How many cars in total were for sale at any time?

see answer
6

# Letter for the Pentagon

Which letter replaces the question mark?

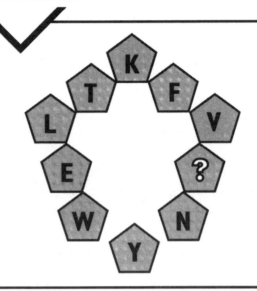

see answer
23

# Dots or Not?

What comes next?

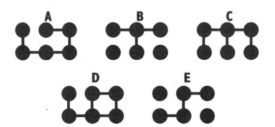

see answer
15

# Route March

A man decides to walk from point A to point B, and for extra exercise to do it by walking round every possible route on the way. Every possible journey from A to B must travel over every part of the four circular paths, i.e. a journey from A to B in a straight line is not permitted because it does not take in any of the circular paths. How many different possible ways are there to walk from A to B by travelling over every route? He does not travel over any part of the route twice, but inevitably arrives at the same point on the straight A/B route more than once on his travels.

*see answer*

**18**

# Puzzling Pentagon

Reading top to bottom and left to right:

If 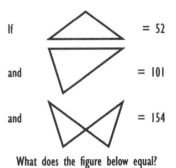 = 52

and = 101

and = 154

What does the figure below equal?

*see answer*

**7**

# Squared Anology

 is to

as  was to:

A    B    C    D    E

*see answer*

**24**

# Circular Oddity

Which is the odd one out?

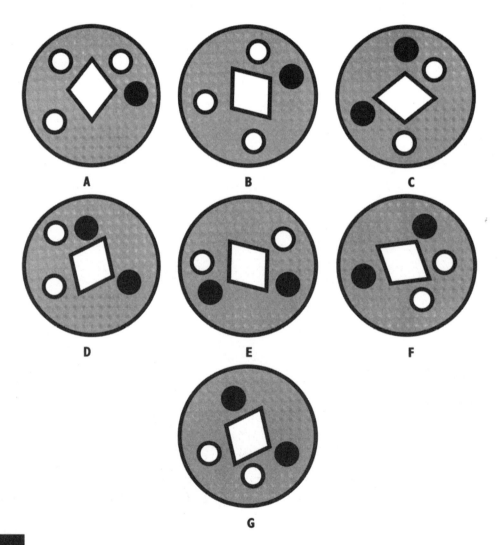

A

B

C

D

E

F

G

see answer

17

128

# The End of the Line

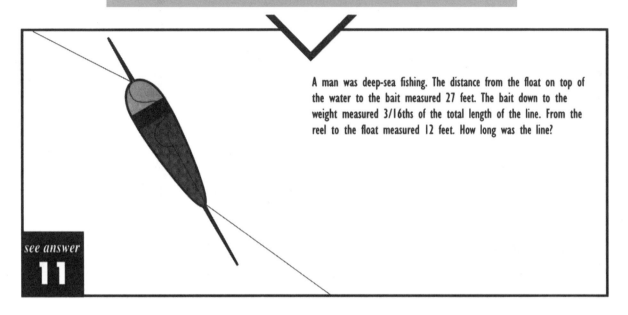

A man was deep-sea fishing. The distance from the float on top of the water to the bait measured 27 feet. The bait down to the weight measured 3/16ths of the total length of the line. From the reel to the float measured 12 feet. How long was the line?

see answer
11

# Bank Raid

I was investigating a bank raid and I wanted to know the letter denoting the year of the getaway car's registration. I asked six witnesses. They answered:

A said, "The letter was made up of straight lines only"
B said, "The letter was a vowel"
C said, "The letter was A, D, or E"
D said, "The letter was G, H, or J"
E said, "The letter was L, M, or N"
F said, "The letter was A, B, or C"

Only four of them had told the truth. Which letter was it?
(The letter could only be A, B, C, D, E, F, G, H, J, K, L, M, N, or P)

see answer
22

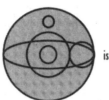 is to

as  is to

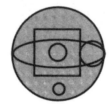

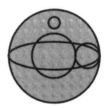

**A**

**B**

**C**

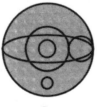

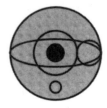

**D**

**E**

see answer
**20**

# Missing Link

From the information given, fill in the missing numbers.

The link between the numbers in each line is the same.

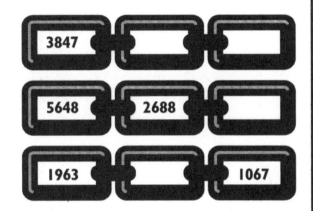

see answer
16

# Depleted Octagon

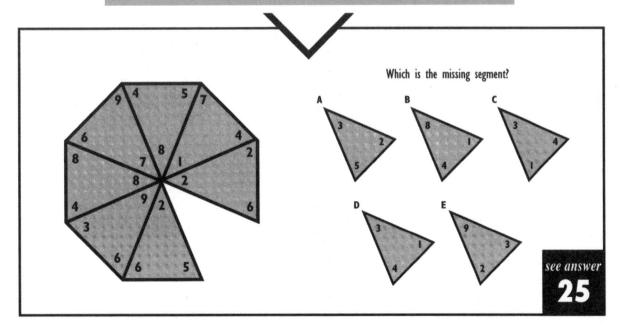

Which is the missing segment?

see answer
25

# Oval Numbers

What number replaces the question mark?

see answer
34

# Cuboid Collection

Which three of these five figures can be fitted together to form a cuboid?

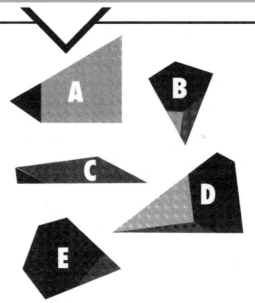

see answer
27

# Symbol Hub

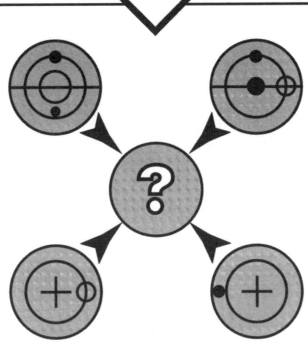

Each line and symbol which appears in the four outer circles is transferred to the centre circle according to these rules:

If a line or symbol occurs in the outer circles;

**once** – it is transferred, **twice** – it is possibly transferred,

**three times** – it is transferred, **four times** – it is not transferred.

Which of the circles A to E should replace the question mark in the diagram?

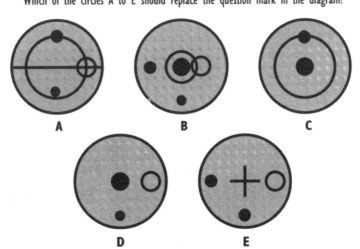

see answer
38

## Shapescape

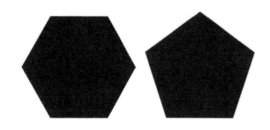

Logically what shape should replace the question mark?

see answer
44

## Magic Hexagon

Distribute the remaining numbers 1 to 12 around the nodes so that each of the six lines of four numbers add up to 26.

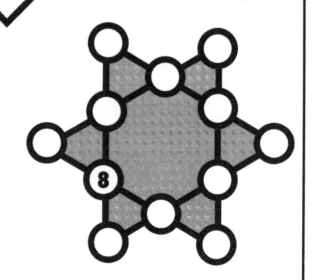

see answer
29

# Central Symbol

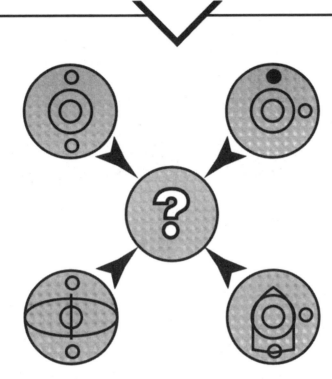

Each line and symbol which appears in the four outer circles is transferred to the centre circle according to these rules:

If a line or symbol occurs in the outer circles;

**once** – it is transferred, **twice** – it is possibly transferred,

**three times** – it is transferred, **four times** – it is not transferred.

Which of the circles A to E should replace the question mark in the diagram?

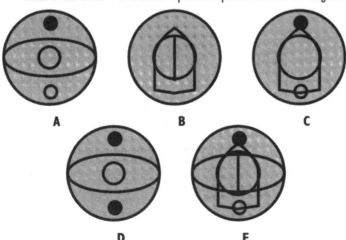

A  B  C

D  E

see answer
**32**

Each of the nine squares in the grid marked 1A to 3C, should incorporate all the lines and symbols which are shown in the squares of the same letter and number at the top and far left of the grid. For example, 3C should incorporate all the lines and symbols that are in boxes 3 and C. One of the squares is incorrect, which one?

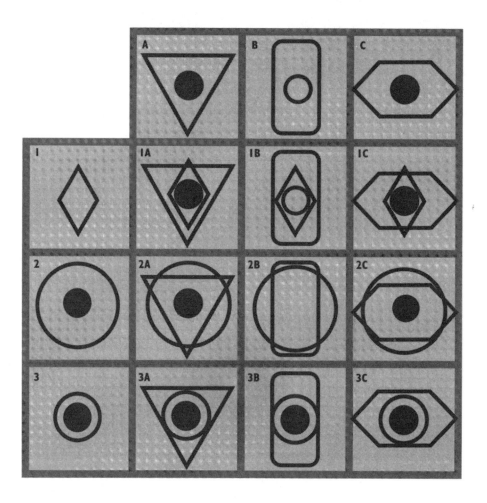

*see answer*

**37**

# Tile Trial

Which tile from A to F replaces the question mark?

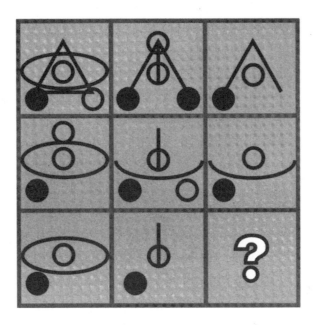

A       B       C

D       E       F

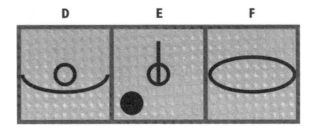

see answer
41

# Square Bashing

The numbers in the grid below relate to their position. What number should replace the question mark?

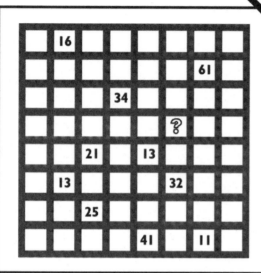

| | | | | | | | |
|---|---|---|---|---|---|---|---|
| | 16 | | | | | | |
| | | | | | | 61 | |
| | | | 34 | | | | |
| | | | | | ? | | |
| | | 21 | | 13 | | | |
| | 13 | | | | 32 | | |
| | | 25 | | | | | |
| | | | | 41 | | 11 | |

see answer
26

# Number Stack

Which set of three numbers logically replaces the question mark?

| | | | |
|---|---|---|---|
| 71 | | | |
| 27 | 72 | | |
| 19 | 91 | 27 | |
| 34 | 43 | 34 | ? |
| 42 | 24 | 42 | ? |
| 11 | 11 | 11 | ? |

| A | B | C | D |
|---|---|---|---|
| 27 | 34 | 27 | 72 |
| 42 | 42 | 34 | 43 |
| 11 | 11 | 11 | 11 |

see answer
36

# Hexagonal Quest

Which tile from A to E replaces the question mark?

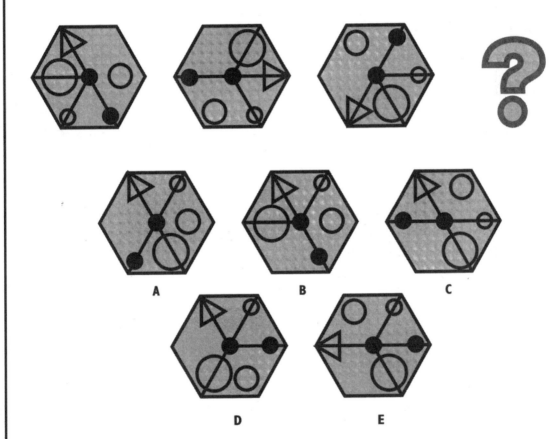

A

B

C

D

E

see answer
42

# Time in Reserve

A game for 36 players lasts for 45 minutes. There are four reserves who alternate equally with each player. This means that all players, including reserves, are on the pitch for the same length of time. How long is that?

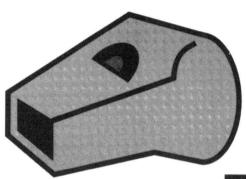

*see answer*
**28**

# Productivity Problem

A company offers to increase wages providing staff increase production by 2.4% per week. If the company works a 6-day week one week and a 5-day week the next, and continues this rota, by how much per day over a monthly period must the workforce increase production to achieve the target?

*see answer*
**48**

# Logical Jigsaw

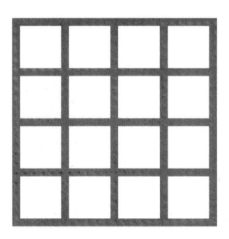

Each row has numbers going +3, −2, +3, and each column has numbers going −3, +2, −3. Place all the numbered pieces into the grid to fulfil these criteria.

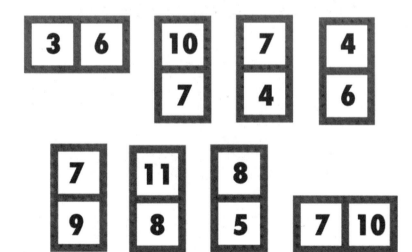

see answer 31

# Sounds Spooked

There were five staff at five different places, each heard a different sound from a different spook. From the following information can you find who heard which sound, from whom, and where?

The wraith was heard by the cook, the gamekeeper lived in the castle.

The cook heard knocking, the butler heard a scream.

The ghoul was in the monastery, the spirit was in the attic.

The phantom was heard by the gamekeeper, the gardener heard a shriek.

The ghost made a scream, knocking was heard in the tower.

The butler was in the manor, the maid was in the attic.

The maid heard a whistle, the phantom made a gurgle.

The gurgle came from the castle.

|  |  | PLACE | | | | | STAFF | | | | | NOISE | | | | |
|---|---|---|---|---|---|---|---|---|---|---|---|---|---|---|---|---|
|  |  | CASTLE | MONASTERY | MANOR | ATTIC | TOWER | GARDENER | BUTLER | COOK | MAID | GAME KEEPER | WHISTLE | GURGLE | KNOCKING | SHRIEK | SCREAM |
| SPOOK | GHOST | | | | | | | | | | | | | | | |
| | WRAITH | | | | | | | | | | | | | | | |
| | PHANTOM | | | | | | | | | | | | | | | |
| | SPIRIT | | | | | | | | | | | | | | | |
| | GHOUL | | | | | | | | | | | | | | | |
| NOISE | WHISTLE | | | | | | | | | | | | | | | |
| | GURGLE | | | | | | | | | | | | | | | |
| | KNOCKING | | | | | | | | | | | | | | | |
| | SHRIEK | | | | | | | | | | | | | | | |
| | SCREAM | | | | | | | | | | | | | | | |
| STAFF | GARDENER | | | | | | | | | | | | | | | |
| | BUTLER | | | | | | | | | | | | | | | |
| | COOK | | | | | | | | | | | | | | | |
| | MAID | | | | | | | | | | | | | | | |
| | GAME KEEPER | | | | | | | | | | | | | | | |

| SPOOKS | PLACE | STAFF | NOISE |
|---|---|---|---|
| | | | |
| | | | |
| | | | |
| | | | |
| | | | |

see answer
40

# Show Certainty

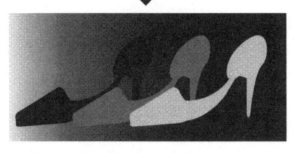

In a darkened room there are:
25 pairs of black shoes, 23 pairs of white shoes and
21 pairs of red shoes.
Each pair is of a slightly different design.
How many shoes must you select to be certain of obtaining a matching pair?

see answer
46

# Hexagonal Numbers

What number replaces the question mark?

16  17
10  **33**  22
8   4

14  16
17  **15**  19
8   9

30  28
7  **?**  14
21  20

see answer
35

# Circular Relative

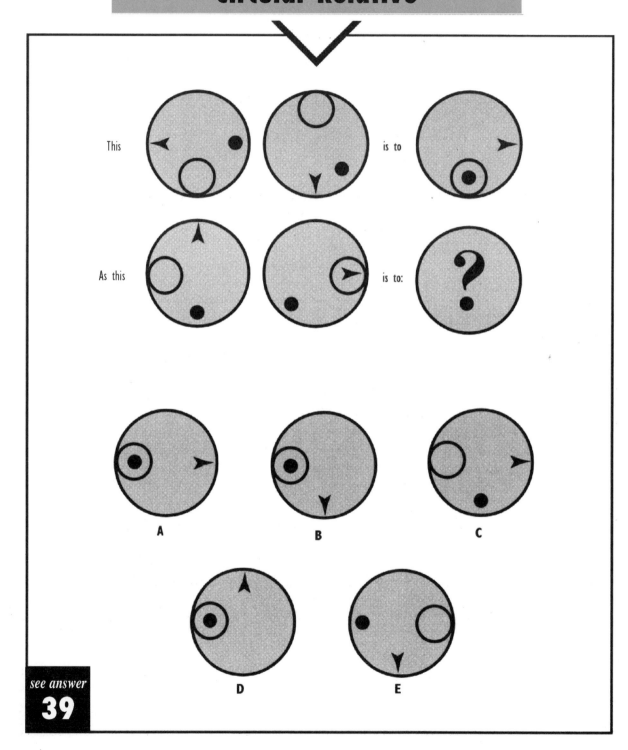

This □ □ is to □

As this □ □ is to: **?**

A

B

C

D

E

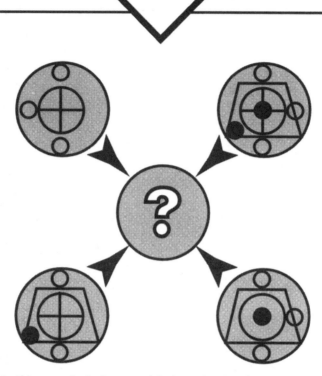

Each line and symbol which appears in the four outer circles is transferred to the centre circle according to these rules:

If a line or symbol occurs in the outer circles;

**once** – it is transferred, **twice** – it is possibly transferred,

**three times** – it is transferred, **four times** – it is not transferred.

Which of the circles A to E should replace the question mark in the diagram?

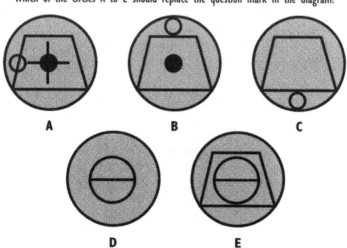

A          B          C

D          E

*see answer*

**43**

145

Which hexagon replaces the question mark?

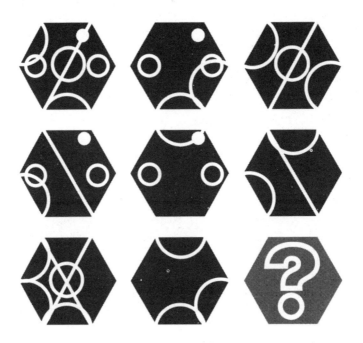

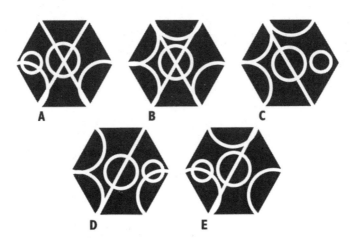

see answer
33

# Casino

The casino rolls two standard six-sided dice and the player rolls one standard six-sided die. The player wins if she rolls a number on her die which is between the two numbers that the casino rolls. Anything else, including ties, the house wins. What is the probability of the player winning?

see answer
**30**

# Deduction Sequence

# 11, 24, 39, 416, 525, ?

What number comes next?

see answer
**45**

# Octagon Missing

Which is the missing segment?

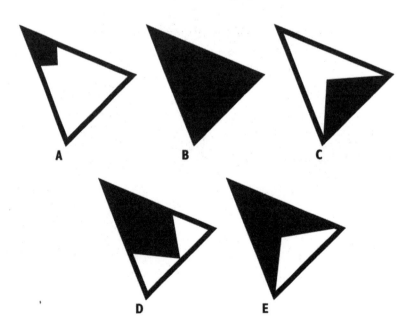

A        B        C

D        E

see answer
47

# Circular Pyramid

Which circle from A to E replaces the question mark, continuing the sequence?

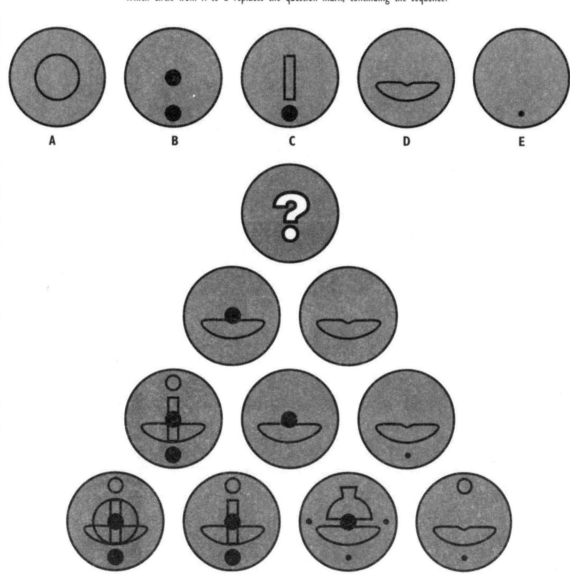

A

B

C

D

E

see answer
63

# Circular Squares

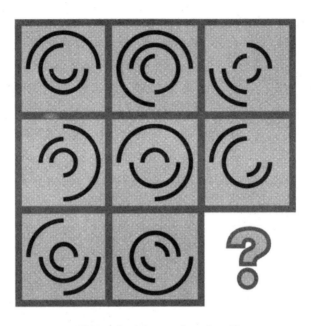

Which option below completes the grid?

**A**

**B**

**C**

**D**

**E**

see answer

**55**

150

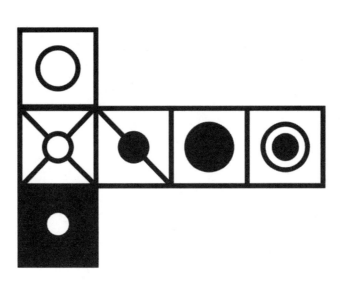

When the above shape is folded to form a cube, just one of the following can be produced. Which one?

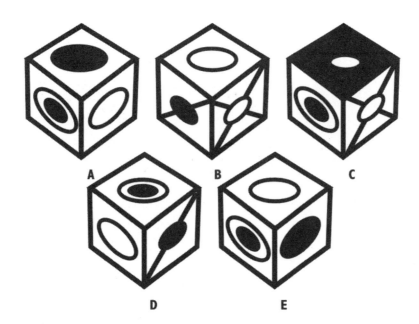

A

B

C

D

E

see answer
49

151

# Symbol Point

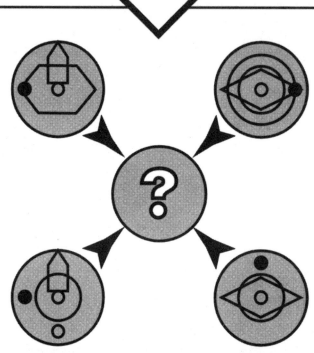

Each line and symbol which appears in the four outer circles is transferred to the centre circle according to these rules:
If a line or symbol occurs in the outer circles;
**once** – it is transferred, **twice** – it is possibly transferred,
**three times** – it is transferred, **four times** – it is not transferred.
Which of the circles A to E should replace the question mark in the diagram?

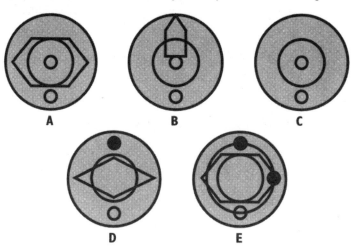

A     B     C

D     E

*see answer*
**58**

152

# Time Teaser

How many minutes is it before 12 noon if, twenty minutes ago, it was four times as many minutes past 9 a.m. as it is before 12 noon now?

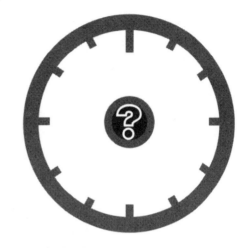

*see answer*
**52**

# Numerous Analogy

## 6895 : 1513 : 28

Which set of numbers below have the same relationship to each other as those above?

A. 6432 : 1862 : 13
B. 6478 : 1312 : 25
C. 5146 : 2804 : 33
D. 7842 : 1410 : 54
E. 9827 : 1709 : 89

*see answer*
**67**

# Average Cricket Problem

A batsman is out for 10 runs which reduces his batting average for the season from 34 to 32. How many runs would he have needed to score to increase his average from 34 to 36?

*see answer*
**56**

# Absent Number

What is the missing number?

*see answer*
**60**

# Circular Square

Which circle from A to E replaces the question mark, continuing the sequence?

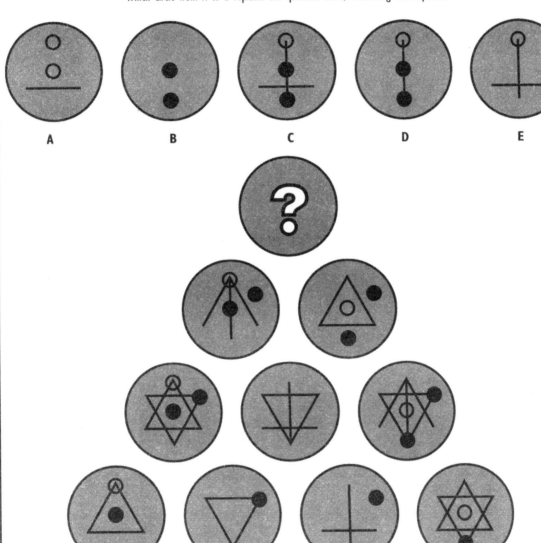

A    B    C    D    E

see answer
53

# Holiday Quintet

Five men and five women, went to five different cities, by five different modes of transport. From the information below, sort out which pair went together to which city and by which mode of transport.

The ferry was used to go to Paris, Alf went with Kate.
Bob went by bus, Sally went to Rome.
Len went by plane, the cruise went to New York.
Mary went to Madrid, Sam went with Judith.
Ben went by train, Fiona went to Berlin.
Alf went to New York, Judith went by ferry.
Kate went on a cruise, Len went to Rome.
Mary went by bus, Ben went to Berlin.

| | | WOMAN | | | | | TRANSPORT | | | | | CITY | | | | |
|---|---|---|---|---|---|---|---|---|---|---|---|---|---|---|---|---|
| | | SALLY | JUDITH | KATE | MARY | FIONA | CRUISE | FERRY | BUS | PLANE | TRAIN | PARIS | BERLIN | MADRID | ROME | NEW YORK |
| MAN | BOB | | | | | | | | | | | | | | | |
| | SAM | | | | | | | | | | | | | | | |
| | ALF | | | | | | | | | | | | | | | |
| | LEN | | | | | | | | | | | | | | | |
| | BEN | | | | | | | | | | | | | | | |
| CITY | PARIS | | | | | | | | | | | | | | | |
| | BERLIN | | | | | | | | | | | | | | | |
| | MADRID | | | | | | | | | | | | | | | |
| | ROME | | | | | | | | | | | | | | | |
| | NEW YORK | | | | | | | | | | | | | | | |
| TRANSPORT | CRUISE | | | | | | | | | | | | | | | |
| | FERRY | | | | | | | | | | | | | | | |
| | BUS | | | | | | | | | | | | | | | |
| | PLANE | | | | | | | | | | | | | | | |
| | TRAIN | | | | | | | | | | | | | | | |

| MALE | FEMALE | CITY | TRANSPORT |
|---|---|---|---|
| | | | |
| | | | |
| | | | |
| | | | |
| | | | |

see answer
62

# Crazy Columns

What are the missing numbers?

| | | | |
|---|---|---|---|
| 2 | 3 | 1 | 0 |
| 4 | 1 | 3 | 6 |
| 5 | 9 |  | 1 |
| 16 | 2 | 7 | 12 |
|  | 15 | 25 | 3 |
| 36 | 5 | 13 | 18 |
| 17 |  | 49 | 8 |
| 64 | 13 | 19 | 24 |
| 23 | 27 | 81 |  |
| 100 | 34 | 29 | 30 |
| 31 | 33 | 121 | 55 |
| 144 | 89 | 37 | 36 |

see answer
66

# Quarters

Divide the grid into four equal segments. Each segment must contain the same symbols, that is three of each: triangle, circle, diamond.

see answer
69

# Bowling Dilemma

Two players have identical bowling balls. The bowling ball of the first travels the full length of the lane at an average speed of 40 mph, while the ball of the second player does the same distance at an average of 35 mph. Which ball goes through the greatest number of revolutions?

see answer
65

# Numerical Palindrome

# 31, 3, 5, 11, 23, 13

These numbers can be transformed into a palindromic number sequence, that is, one that reads the same backwards as forwards if the same formula is applied to each of them. What is the formula and what are the resulting numbers?

see answer
54

## Birthday Secret

I wanted to know the birthday of a work colleague, I knew it was during February but did not know the actual date. I asked six other colleagues. This is how they answered.

A said, "It is an odd number."
B said, "It is a prime number."
C said, "It is between 6 and 16."
D said, "It is before 10."
E said, "It is a double-digit number."
F said, "It is between 8 and 12."

One of them had lied. What was the date?
(1 is not considered to be a prime number)

see answer
50

## ABC

Twenty-six cards, each featuring a different letter of the alphabet are shuffled and then placed face down on a table and turned over at random one by one. What are the chances that the first three cards to be turned over are A, B, C, in that order?

see answer
57

# Spots Before the Eyes

Which is the odd spot out?

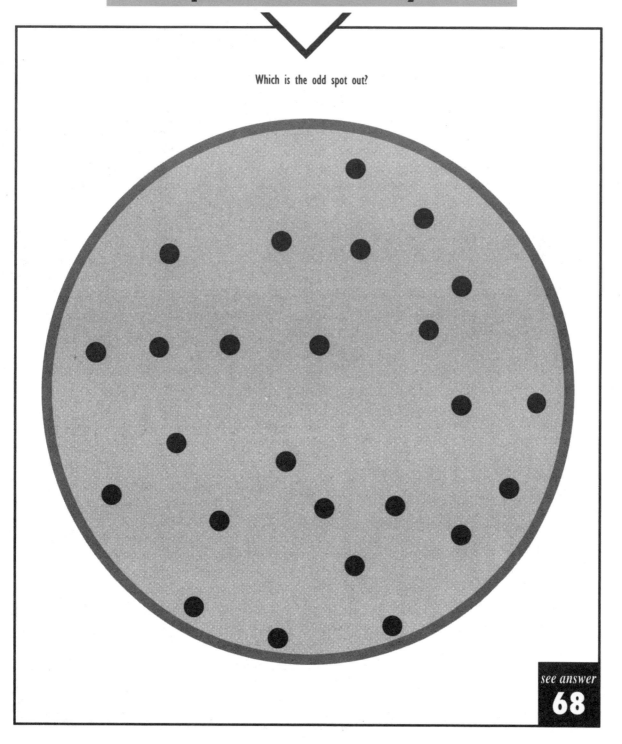

see answer
**68**

## Sneaky Sequence

# 18, 27, 64, 12, 52, 16, 34, 35, ?

What number continues the sequence?
Clue: the commas are there to confuse you!

see answer
59

## Number Jump

| 4 | 7 | 8 | 5 |
|----|----|----|----|
| 9 | 11 | 4 | 2 |
| 12 | 6 | ? | 9 |
| 7 | 2 | 7 | 12 |

What number should replace the question mark?

see answer
51

# Dotty Pentagon

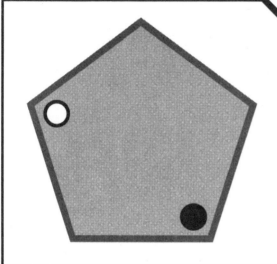

The black dot moves two corners counter-clockwise at each stage and the white dot moves three corners clockwise at each stage. In how many stages will they be together in the same corner?

see answer
61

# Fractional Feature

What feature is shared by three out of the following four fractions?

$$\frac{16}{64} \quad \frac{14}{42} \quad \frac{19}{95} \quad \frac{26}{65}$$

see answer
64

In a variation of Russian roulette, the gun, a six-shooter revolver, has two bullets in a row in two of the chambers. The barrel is spun once and each of the two players points the gun at their head in turn and pulls the trigger. If the gun fails to go off the gun is passed to the next player who repeats the process. In this deadly game the decision about who goes first is made by the toss of a coin, i.e. heads you go first, tails the other player shoots first. Would you rather be the first or second to shoot?

*see answer*
**70**

Ignoring rotations and reflections it is possible to dissect a pentagon into (three) triangles in only one way using the rule that all lines drawwn must travel from corner of th pentagon and must not bisect each other.

Using the same rule, how many ways can a heptagon be dissected, and how many triangles are produced?

*see answer*
**74**

# Pentagonal Pyramid

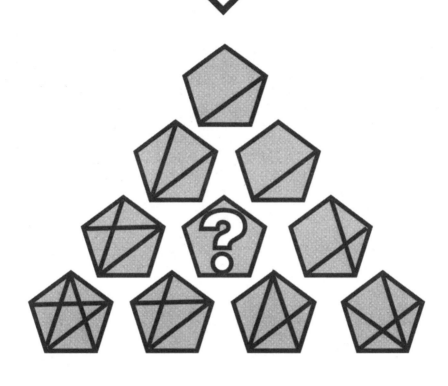

Which is the missing pentagon?

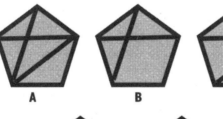

A        B        C

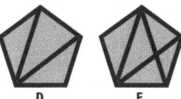

D        E

*see answer*
**80**

## Strange Sums

If half of 9 is 4, and half of 11 is 6, and half of 12 is 7, what is half of 13?

*see answer*
**73**

## Multiplication Mystery

 ×  =

Substitute numbers for symbols in this multiplication problem. Each symbol represents a different number from 1 to 9. There are two possible answers in which the following symbols represent the same number in both:

*see answer*
**81**

166

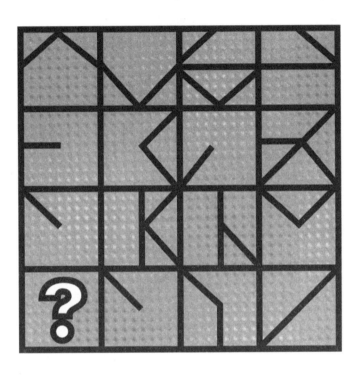

What should replace the question mark?

A

B

C

D

E

*see answer*

**75**

## Teeming with Triangles

How many triangles are in this drawing?

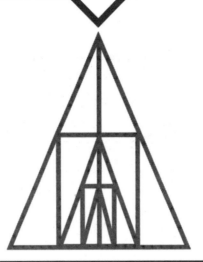

see answer
87

## Bracket Gap

428 (64105) 375
293 (5456) 472
734 (        ) 289

What number is missing from the third bracket?

see answer
79

# Dots Spot

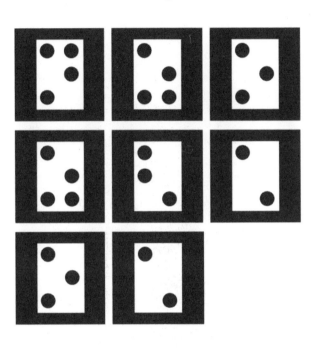

Which is the missing square

A

B

C

D

E

see answer
82

169

# Twins and Triplets

Two friends, Susan and Anne, each have five children, twins and triplets.

Susan's twins are older than her triplets, whereas Anne's triplets are older than her twins. Susan remarked recently that the sum of the ages of her children was equal to the product of their ages.

Anne replied that the same was true of her children.

How old are Susan and Anne's children?

*see answer*

**83**

# Dots Dilemma

What continues this sequence?

A      B      C      D      E

*see answer*

**72**

# Tricky Triangles

What number should replace the question mark?

see answer
76

# Melting Hex

Which option continues the above sequence?

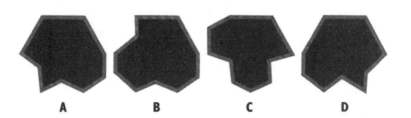

A        B        C        D

see answer
91

Which circle fits the missing space?

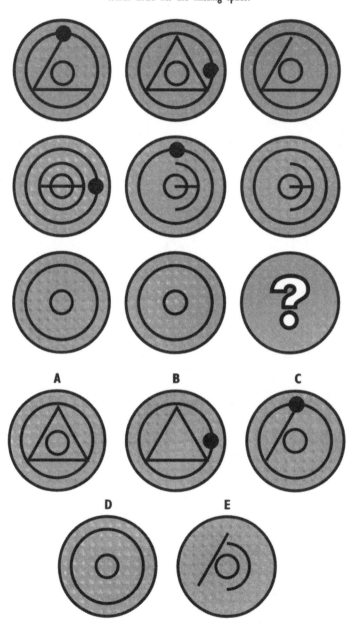

A

B

C

D

E

see answer
92

# Cunning Connections

Insert the numbers 0–8 in the circles, so that for any particular circle, the sum of the numbers in the circles connected directly to it, equals the value corresponding to the number in the circle in the following list:

0 = 16, 1 = 13, 2 = 6, 3 = 0, 4 = 6, 5 = 7, 6 = 7, 7 = 14, 8 = 7.

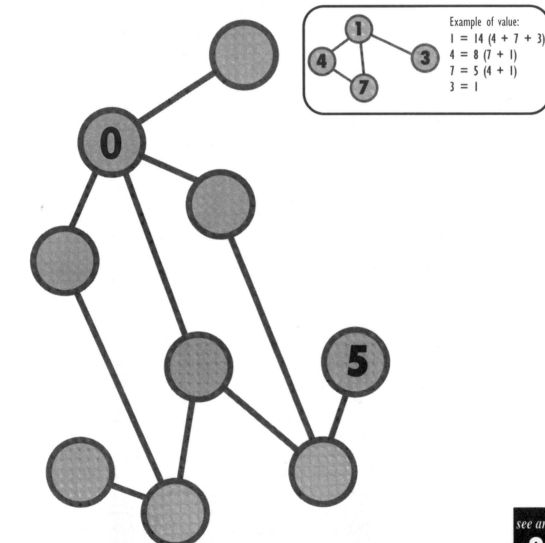

Example of value:
1 = 14 (4 + 7 + 3)
4 = 8 (7 + 1)
7 = 5 (4 + 1)
3 = 1

see answer
84

# Griddle

What number should replace the question mark?

| | | | 41 | | |
|---|---|---|---|---|---|
| | 22 | | | | |
| | | | | 53 | |
| 14 | | | | | |
| | | 35 | | | |
| | | | | | ? |

see answer
89

# Brainstrain

What symbol is two to the left of the symbol immediately to the left of the symbol four to the right of the symbol which is midway between the symbol two to the left of the

and the symbol three to the left of the

see answer
77

# Treasure Trail

The treasure is on the square marked 'T'. To reach it you have to find the starting square which will direct you to every one of the 35 remaining squares, only once each, before arriving at the treasure. '1S' means one square south, '4E' means four squares east, etc.

| | | | | | |
|---|---|---|---|---|---|
| 5 E 1 S | 4 E 3 S | 2 W 4 S | 2 S 2 W | 3 W 4 S | 1 S 2 W |
| 5 E 1 N | 2 E 4 S | 1 S 3 E | 1 S 1 W | 4 W 1 N | 2 W 1 S |
| 3 E 2 S | 1 S 1 W | 2 S 3 E | 1 N 3 W | 3 S 4 W | 2 N 1 W |
| 3 N 1 E | 3 E 2 S | 2 S 1 W | **T** | 2 W 2 S | 1 W 1 S |
| 5 E 1 S | 1 W 2 N | 4 N 1 E | 2 N 1 E | 1 N 2 W | 3 N 4 W |
| 1 N 2 E | 1 E 5 N | 2 N 1 W | 1 E 2 N | 1 W 2 N | 1 W 4 N |

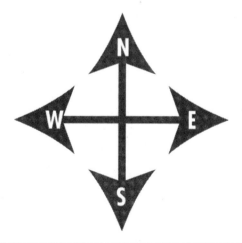

see answer
86

## Get the Needle

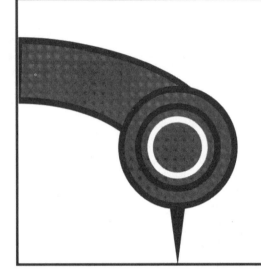

A gramophone record is 12 inches is diameter. It has a 5 inch diameter middle and a ½ inch outer border. The remaining playing surface has 100 grooves to the inch. How far does the needle travel during the playing time of one side?

*see answer*
**78**

## Strange Journey

On a particular journey I travel the first half of the distance at 25 mph. How fast would I need to travel to average 50 mph for the complete trip?

*see answer*
**88**

# Pyramidal Logic

Which symbols should replace the question marks?

see answer
71

# Horse Race

There were five horses in a race, each with a differently named jockey. Each jockey had a different shirt and cap. Find which jockey rode which horse with which shirt and cap, from the following information.

Mistral's jockey had a red cap, Roger had a striped shirt.

Trevor had a pink cap, Genoa's jockey had stars on his shirt.

Arcturus was ridden by Archie.

The plain shirt goes with a blue cap.

Maurice had a yellow cap, Brian rides Lullaby.

The jockey riding Palisade has diamonds on his shirt.

Hooped shirt goes with a mauve cap.

Genoa's rider has a yellow cap.

Mistral's rider has a striped shirt, Archie has a blue cap.

Maurice has a shirt with stars, Brian's shirt is hooped.

|  |  | ARCHIE | BRIAN | ROGER | MAURICE | TREVOR | MAUVE | YELLOW | RED | BLUE | PINK | PLAIN | DIAMONDS | STRIPED | STARS | HOOPED |
|---|---|---|---|---|---|---|---|---|---|---|---|---|---|---|---|---|
| HORSE | ARCTURUS | | | | | | | | | | | | | | | |
| | LULLABY | | | | | | | | | | | | | | | |
| | MISTRAL | | | | | | | | | | | | | | | |
| | GENOA | | | | | | | | | | | | | | | |
| | PALISADE | | | | | | | | | | | | | | | |
| SHIRT | PLAIN | | | | | | | | | | | | | | | |
| | DIAMONDS | | | | | | | | | | | | | | | |
| | STRIPED | | | | | | | | | | | | | | | |
| | STARS | | | | | | | | | | | | | | | |
| | HOOPED | | | | | | | | | | | | | | | |
| CAP | MAUVE | | | | | | | | | | | | | | | |
| | YELLOW | | | | | | | | | | | | | | | |
| | RED | | | | | | | | | | | | | | | |
| | BLUE | | | | | | | | | | | | | | | |
| | PINK | | | | | | | | | | | | | | | |

| HORSE | JOCKEY | CAP | SHIRT |
|---|---|---|---|
| | | | |
| | | | |
| | | | |
| | | | |
| | | | |

see answer
90

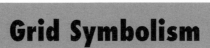

# Grid Symbolism

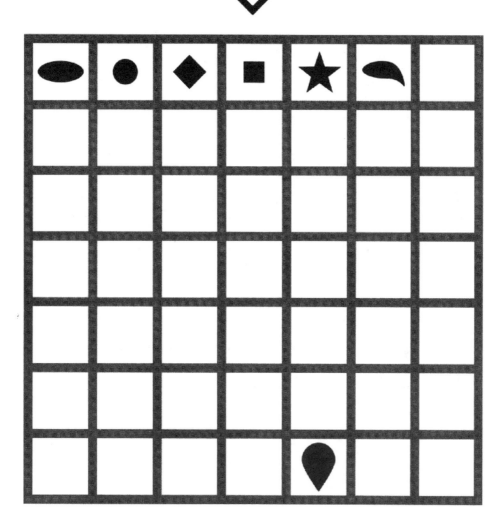

Using only the symbols shown, complete the rest of the grid in such a way that every square is filled and no symbol is repeated on the same horizontal, diagonal or vertical line. Note that diagonal means every diagonal line, not just corner to corner!

see answer
85

# Hex in Space

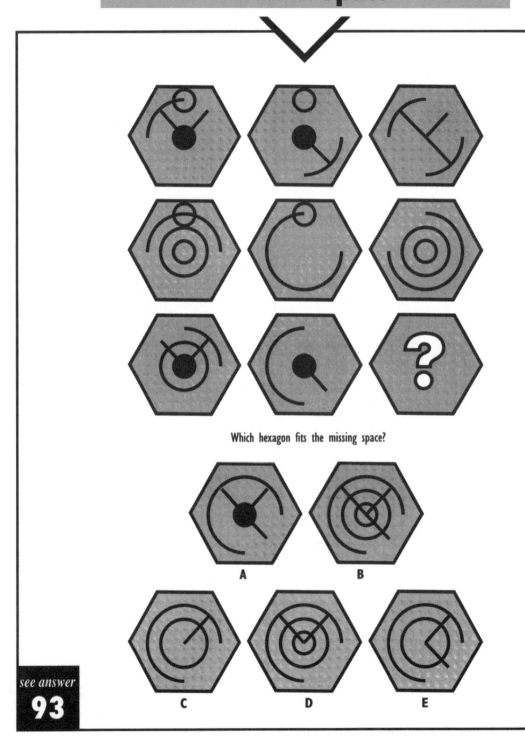

Which hexagon fits the missing space?

A

B

C

D

E

see answer
93

# Bridge Building

There are five different types of bridges, each in a different country and with a different name and length. Can you connect the details correctly?

The bridge at Howrah is a cantilever, the Canadian bridge is an arch.

The bridge in India is 457m long, the bridge in Port Mann is 366m long.

The suspension bridge is 1014m long, the truss bridge is in the USA.

The Canadian bridge is not 1014m long, the bridge in Port Mann is an arch.

Gladesville is in Australia, the concrete bridge is 305m long.

The American bridge is 488m long, Salazar is in Portugal.

The Cooper bridge is a truss, the bridge in Gladesville is not a suspension bridge.

|  | | COUNTRY | | | | | LENGTH | | | | | TYPE | | | | |
|---|---|---|---|---|---|---|---|---|---|---|---|---|---|---|---|---|
|  |  | INDIA | U.S.A. | CANADA | AUSTRALIA | PORTUGAL | 488m | 366m | 1014m | 305m | 457m | ARCH | TRUSS | CONCRETE | CANTILEVER | SUSPENSION |
| BRIDGE | PORT MANN | | | | | | | | | | | | | | | |
|  | SALAZAR | | | | | | | | | | | | | | | |
|  | GLADESVILLE | | | | | | | | | | | | | | | |
|  | HOWRAH | | | | | | | | | | | | | | | |
|  | COOPER | | | | | | | | | | | | | | | |
| TYPE | ARCH | | | | | | | | | | | | | | | |
|  | TRUSS | | | | | | | | | | | | | | | |
|  | CONCRETE | | | | | | | | | | | | | | | |
|  | CANTILEVER | | | | | | | | | | | | | | | |
|  | SUSPENSION | | | | | | | | | | | | | | | |
| LENGTH | 488m | | | | | | | | | | | | | | | |
|  | 366m | | | | | | | | | | | | | | | |
|  | 1014m | | | | | | | | | | | | | | | |
|  | 305m | | | | | | | | | | | | | | | |
|  | 457m | | | | | | | | | | | | | | | |

| NAME | COUNTRY | LENGTH | TYPE |
|---|---|---|---|
|  |  |  |  |
|  |  |  |  |
|  |  |  |  |
|  |  |  |  |
|  |  |  |  |

see answer

181

# Slippery Symbols

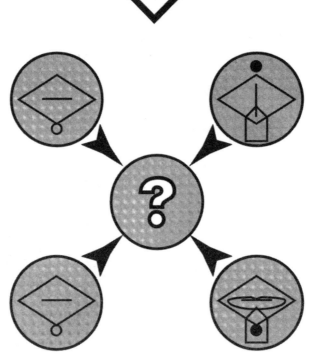

Each line and symbol which appears in the four outer circles is transferred to the centre circle according to these rules:
If a line or symbol occurs in the outer circles;
**once** – it is transferred, **twice** – it is possibly transferred,
**three times** – it is transferred, **four times** – it is not transferred.
Which of the circles A to E should replace the question mark in the diagram?

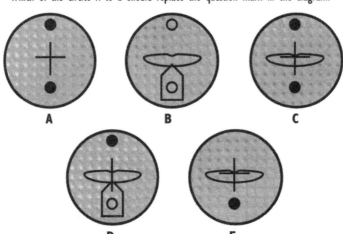

see answer
102

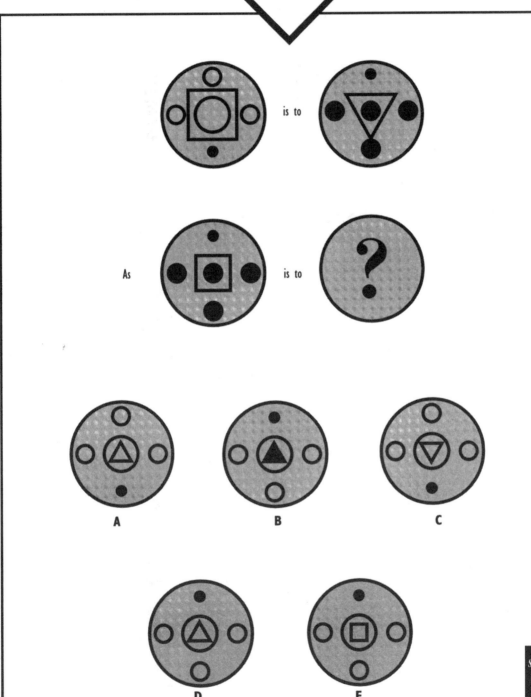

is to

As

is to

?

A

B

C

D

E

see answer
108

## Cagey Croupiers

A casino manager wished to check the accuracy of the roulette wheel so he asked his six croupiers to let him know which number came up most frequently during the days operation. The wheel ran from 1 to 36, and had no zero or double zero.

A said, "It was even."
B said, "It was prime."
C said, "It had double digits."
D said, "It was between 3 and 18."
E said, "It had at least one digit number 1 in it."
F said, "It was 15, 25, 31 or 35."

Half the croupiers had lied. Which number was it?
(1 is not considered to be a prime number)

*see answer*
**103**

## Shape Shift

Which shape should replace the question mark? Choose from:

*see answer*
**96**

A     B     C     D

Logically, which tile from A–F should replace the question mark?

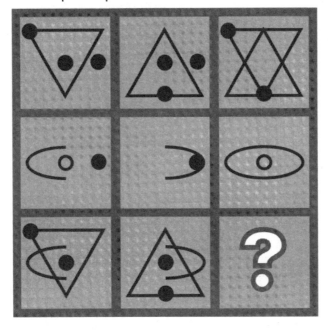

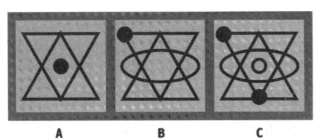

A          B          C

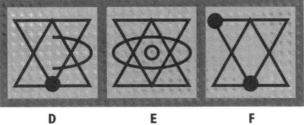

D          E          F

see answer
**100**

# Hexagons in the Round

Which symbols should go in the ? hexagon, and where should they be placed?

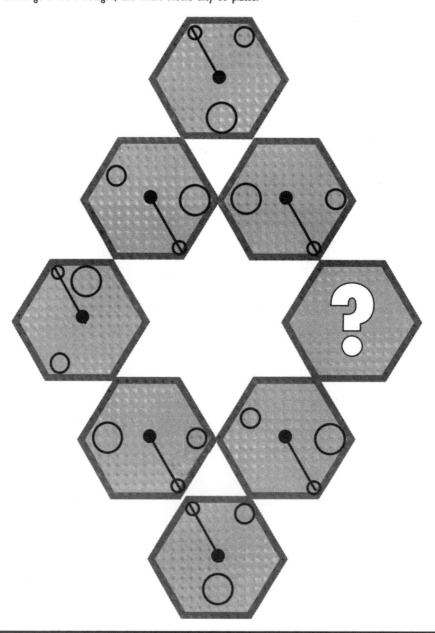

# Shifty Symbols

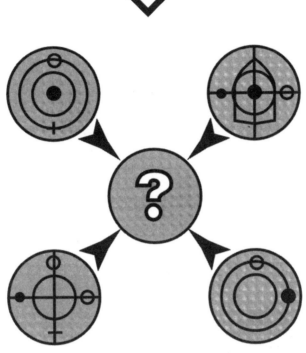

Each line and symbol which appears in the four outer circles is transferred to the centre circle according to these rules:

If a line or symbol occurs in the outer circles;

**once** – it is transferred, **twice** – it is possibly transferred,

**three times** – it is transferred, **four times** – it is not transferred.

Which of the circles A to E should replace the question mark in the diagram?

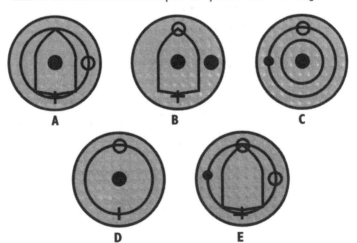

A    B    C

D    E

see answer
**94**

187

# Bar Queue

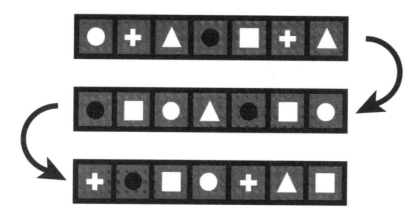

Which option continues the above sequence?

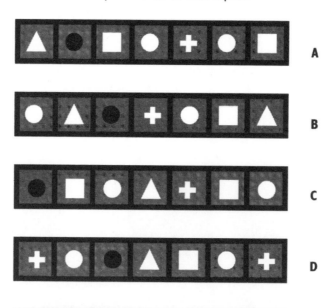

A

B

C

D

E

*see answer* **105**

# Hexagonal Roll

Which hexagon continues the sequence?

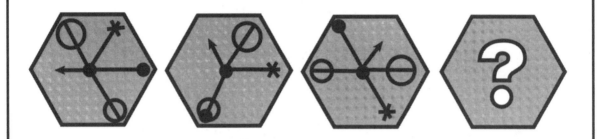

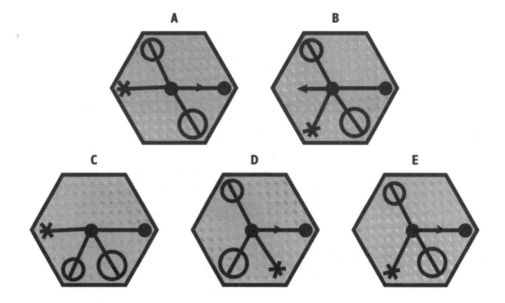

see answer
109

## Natural System

Four people from an undiscovered island were discussing numbers in their system.

A said 18 was a prime number. B said 7 x 8 = 62. C said 36 was a prime number. D said 64 was divisible by 4.

Only two people were telling the truth. What was the base of their number system?

*see answer*
**95**

## Full Family

Brian was one child in a large family, the other children were called Toni, Sam, Alex, Viv and Anthony. When they all grew up Brian had a sister-in-law called Sarah. From this information, who is Sarah married to?

*see answer*
**98**

Which hexagon below continues the sequence?

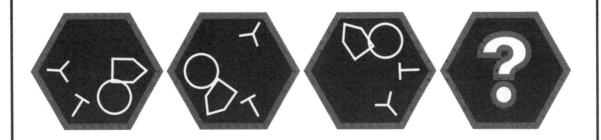

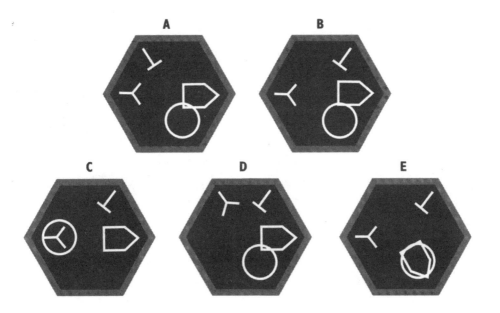

see answer
101

# Golf Poser

Two professional golfers were playing a challenge game between themselves to see who was the better player. Player A scored 69 and player B scored 72, yet player B was the winner.

How could that be? (They were not playing match play).

*see answer*
**97**

# Number Flags

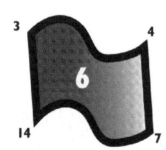

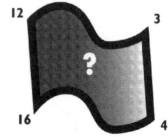

3    4     7    9     12    3
   6           21           ?
14    7     9    3     16    4

*see answer*
**106**

What number should replace the question mark?

# Combination Golf

Five different golfers each played at a different course, using a different golf club, at a different hole number. From the following information work out who was where and with what.

Fred was at hole 14, the driver was used at Wentworth.
Colin used the wedge, the wood was used at hole 12.
At Troon it was hole 10, Joe was at St Anne's.
The driver was used at hole 8, the putter was used at Carnoustie.
Alec used the iron, at Gleneagles it was hole 6.
Fred was at Carnoustie, Bill was at hole 8.
Alec was at the 10th hole, Colin was at hole 6.

|  |  | WENTWORTH | GLENEAGLES | ST. ANNES | TROON | CARNOUSTIE | WOOD | DRIVER | PUTTER | WEDGE | IRON | 6 | 8 | 10 | 12 | 14 |
|---|---|---|---|---|---|---|---|---|---|---|---|---|---|---|---|---|
| GOLFER | JOE |  |  |  |  |  |  |  |  |  |  |  |  |  |  |  |
|  | BILL |  |  |  |  |  |  |  |  |  |  |  |  |  |  |  |
|  | FRED |  |  |  |  |  |  |  |  |  |  |  |  |  |  |  |
|  | ALEC |  |  |  |  |  |  |  |  |  |  |  |  |  |  |  |
|  | COLIN |  |  |  |  |  |  |  |  |  |  |  |  |  |  |  |
| HOLE | 6 |  |  |  |  |  |  |  |  |  |  |  |  |  |  |  |
|  | 8 |  |  |  |  |  |  |  |  |  |  |  |  |  |  |  |
|  | 10 |  |  |  |  |  |  |  |  |  |  |  |  |  |  |  |
|  | 12 |  |  |  |  |  |  |  |  |  |  |  |  |  |  |  |
|  | 14 |  |  |  |  |  |  |  |  |  |  |  |  |  |  |  |
| CLUB | WOOD |  |  |  |  |  |  |  |  |  |  |  |  |  |  |  |
|  | DRIVER |  |  |  |  |  |  |  |  |  |  |  |  |  |  |  |
|  | PUTTER |  |  |  |  |  |  |  |  |  |  |  |  |  |  |  |
|  | WEDGE |  |  |  |  |  |  |  |  |  |  |  |  |  |  |  |
|  | IRON |  |  |  |  |  |  |  |  |  |  |  |  |  |  |  |

| NAME | COURSE | GOLF CLUB | HOLE |
|---|---|---|---|
|  |  |  |  |
|  |  |  |  |
|  |  |  |  |
|  |  |  |  |
|  |  |  |  |

see answer
99

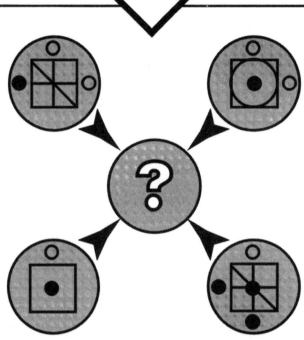

Each line and symbol which appears in the four outer circles is transferred to the centre circle according to these rules:
If a line or symbol occurs in the outer circles;
**once** – it is transferred, **twice** – it is possibly transferred,
**three times** – it is transferred, **four times** – it is not transferred.
Which of the circles A to E should replace the question mark in the diagram?

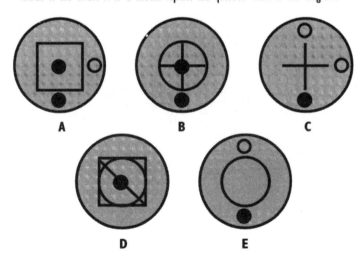

A          B          C

D          E

*see answer*
**110**

**194**

# Magic grid

Where would you place the numbers 1, 3, 10, 16 and 25 to follow a logical pattern?

| 17 | 24 | ? | 8 | 15 |
| 23 | 5 | 7 | 14 | ? |
| 4 | 6 | 13 | 20 | 22 |
| ? | 12 | 19 | 21 | ? |
| 11 | 18 | ? | 2 | 9 |

see answer
**123**

# Town Travel

On a motorway there were four towns

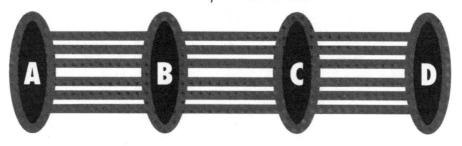

The distance from A to B is 32 m.
The distance from B to C is 3/8ths of the total distance from A to D.
The distance from C to D is 24 m.
What is the distance from A to D?

see answer
**114**

## Quick Detection

A woman goes to her office but finds she has left her purse at home. She returns home to overhear her husband shout "Don't shoot me John" from the bedroom. Rushing into the bedroom she sees her husband unconscious and bleeding on the floor in a corner. Across the room she sees a banker, a refuse collector and an accountant. None of them has a weapon and she doesn't know any of their names. She turns to the refuse collector and says "Why did you shoot him?" How did she know the refuse collector had done it?

*see answer*
**116**

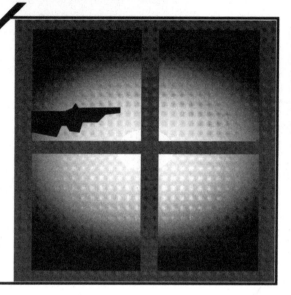

## Decimal Blunders

$$37.6 +$$
$$1921.5 +$$
$$109.4 +$$
$$14.7$$
$$\overline{3876.9}$$

In the following sum only one decimal point is in its correct position. Alter four of the decimal points to make the sum correct. There are two possible answers.

*see answer*
**126**

# Box Clever

Which box continues the sequence?

A                                    B

C                    D                    E

see answer
**113**

# Cube Catch

Which of the options given will fold up to make the cube below?

A

B

C

D

see answer
**117**

# Devilish Dice

The game of dice in a casino involves throwing a pair of dice numbered 1 to 6.

What is the average difference between the two dice?

see answer
111

# Circular Spin

Which number should replace the ?

72
12  36

44
14  82

27
10

see answer
1·20

# Logical Segment

Logically, what number should replace the question mark?

see answer
**121**

# Soccer Scores

Three soccer players were discussing the goals that they had scored during the season.

Player 1:    'I scored 9, I scored two less than player 2, I scored one more than player 3.'
Player 2:    'I did not score the lowest, the difference between my score and player 3 was 3, player 3 scored 12.'
Player 3:    'I scored less than player 1, player 1 scored 10, player 2 scored three more than player 1.'

Each man made one incorrect statement out of three. What were their scores?

see answer
**115**

# Infinite Calculation

Study this infinite series:

## 4 – 4 + 4 – 4 + 4 – 4 + 4 to infinity

Does this equal

## 4 or 0 ?

see answer
**127**

# Match Magic

Construct 5 squares using 24 matches.
(There are at least two solutions)

see answer
**124**

# Box Sympathy

Which of the five boxes below fulfils the criteria of the box above?

**A**

**B**

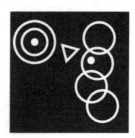

**C**

**D**

**E**

*see answer*
**119**

# Sharp Cards

You have 52 cards marked 1 to 52. If you remove, shuffle and deal the first four cards, what are the chances that they will be in increasing numerical order?

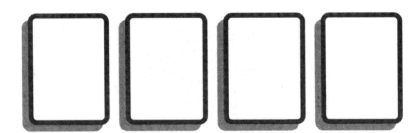

see answer
125

# Triangular Teaser

What number should replace the question mark?

see answer
118

# Calculated Countdown

## 8195 + 1921

If these two numbers total 6879, what do the two numbers below total?

## 8216 + 1909

see answer
112

# Lawn Labour

A lawn is of a certain size. One man can mow it in four hours, another in six hours, a third man in only three hours, while a fourth takes as long as the first. If they all worked on the lawn at the same time at their relative speeds, how long would it take them to finish the job?

see answer
122

## Answer 1
## Suit trick
D.

The arrangement completes every possible grouping of the four suits in which no suit appears in the same position as the first arrangement.

## Answer 2
## Who's dancing?

| NAME | FANCY DRESS | CLOTHES | DANCE |
|------|-------------|---------|-------|
| HENRY | DRACULA | LEGGINGS | BOSSA-NOVA |
| ROBERT | NAPOLEON | JACK BOOTS | CHARLESTON |
| SIMON | DR JEKYLL | FEDORA | JITTERBUG |
| PETER | FRANKENSTEIN | HOMBURG | PALAIS GLIDE |
| MORRIS | SHAKESPEARE | SKULL-CAP | BARN DANCE |

## Answer 3
## Likely coins
i) Certainty.
ii) Even chance.
iii) 3 chances in 13.

The coins can fall 32 possible ways:

1. 5 heads or 5 tails (1 way each).

2. 4 heads or 4 tails (5 ways each).

3. 3 heads or 3 tails (10 ways each).

Therefore, at least three of the coins must always finish heads or tails, so the chances of three heads are even. There are also 26 ways that it cannot and six ways that it can, be four heads. So the chances are 6 in 26, or 3 in 13.

## Answer 4
## Shape up
C.

The white dot moves anti- (counter) clockwise one side, and alternates inside then outside, of the rectangle each time. The triangle does the same except clockwise. The black dot simply moves between the top and bottom of the rectangle each time.

## Answer 5
## Truth chase
10th.
A and D lied.

## Answer 6
## Red car blues
12.
Originally 7 red and 5 blue.

## Answer 7
## Puzzling pentagon
213.
A two-digit number is formed by joining the single numbers at the end of each line of the pentagon, with top to bottom having priority over left to right where there is a clash. These two-digit numbers are then added together to give a value for each section.

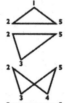

12 + 15 + 25 = 52.

25 + 23 + 53 = 101.

23 + 54 + 24 + 53 = 154.

25 + 23 + 54 + 34 + 24 + 53 = 213.

**Answer 8**
**Keyboard teaser**
K.
It is the only key described three times.

**Answer 9**
**Spelling bee**
116,424

| Boys | | Girls |
|---|---|---|
| $\dfrac{10 \times 9 \times 8 \times 7 \times 6}{1 \times 2 \times 3 \times 4 \times 5}$ | $\times$ | $\dfrac{11 \times 10 \times 9 \times 8 \times 7 \times 6}{1 \times 2 \times 3 \times 4 \times 5 \times 6} = 116,424$ |

**Answer 10**
**Odd shape**
F.
Each of the other figures has an identical pairing but with black and white reversed.

**Answer 11**
**The end of the line**
48 feet.

**Answer 12**
**Democratic digits**
Do the sum 96,284 + 5,392 + 7,845 + 10,219 = 119,740.
Divide this by 4 = 29,935.
29,935 is the number of votes received by the winning candidate.

The second received 29,935 – 5,392 = 24,543.

The third received 29,935 – 7,845 = 22,090.

The fourth received 29,935 – 10,219 = 19,716.

**Answer 13**
**Eleven-tree shuffle**
16 rows at three trees per row.

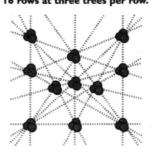

You may find reflections or rotations of this order, but they will be the same.

**Answer 14**
**Circular question**
8.
Left segment ÷ right segment x 5 = top segment.
18 ÷ 15 (1.2) x 5 = 6.
24 ÷ 10 (2.4) x 5 = 12.
8 ÷ 5 = (1.6) x 5 = 8.

**Answer 15**
**Dots or not?**
D.
The shapes are the odd numbers 1, 3, 5 and 7 laid on their side.

## Answer 16
## Missing link

Reading left to right, multiply the left two digits by the right two digits to give the next number.

38 x 47 = 1,786   17 x 86 = 1,462.

56 x 48 = 2,688   26 x 88 = 2,288.

19 x 63 = 1,197   11 x 97 = 1,067.

## Answer 17
## Circular oddity
D.
All the others have an identical companion (A and B, C and E, F and G).

## Answer 18
## Route March
1296 different routes.
There are four different routes bordering the straight route A/B. If there was just one alternative route the man could continue in three ways when reaching the first intersection; left, right or straight ahead. On reaching the second intersection he would have the choice of just two routes, so that altogether there would be six (3 x 2) possible routes. If there were two alternative routes the choice would be $6^2$ (6 x 6) or 36. With four alternative routes the choice is $6^4$ (6 x 6 x 6 x 6) or 1296.

## Answer 19
## Spot the number
6.
The first two numbers in each row and column are divided by 3 or 4, whichever is possible, and the quotients added together to produce the third number. For example (21 ÷ 3) + (8 ÷ 4) = 9.

## Answer 20
## Circular link
D.
The ellipse rotates 90° clockwise; the small circle rotates 180° and becomes a small triangle and vice versa; the middle circle changes shade to black and vice versa; the largest circle becomes a square and vice versa and the circle on the right does not change.

## Answer 21
## Circular connection
A.
The circle changes shade to or from black and all the shapes rotate 180°.

## Answer 22
## Bank raid
A.
D and E did not tell the truth.

**Answer 23**
**Letter for the pentagon**
X.
All the letters are made of straight lines only. The pentagons are in pairs according to their letter's number of straight lines. X is the only non-repeating option.

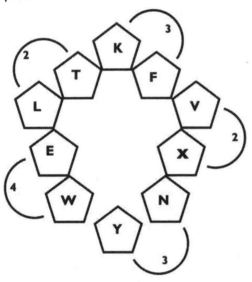

**Answer 24**
**Squared analogy**
E.
The figure turns 90° clockwise and the diamond goes inside the square.

**Answer 25**
**Depleted octagon**
D.
In opposite segments, opposite numbers total 10. That is, if a segment is folded over onto the opposite segment, the two numbers in each corner would add up to 10.

**Answer 26**
**Square bashing**
52.
Each number indicates the number of consecutive blank squares to the left and right of it in the grid. For example a number 41 indicates four consecutive blank squares to the left and one to the right.

**Answer 27**
**Cuboid collection**
A, C and D.

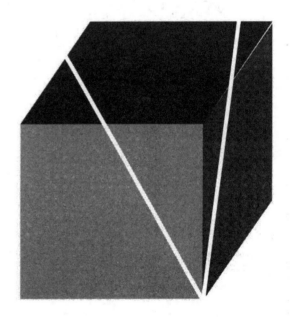

**Answer 28**
**Time in reserve**
40.5 minutes.
Only 36 players can be on the field for 45 minutes, so multiply these two numbers and divide the product by 40 for the number of actual participants.

(45 x 36) [1620] ÷ 40 = 40.5.

## Answer 29
### Magic hexagon

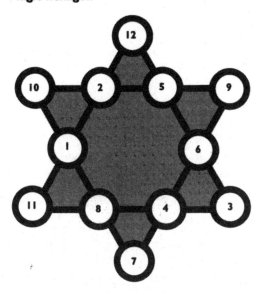

## Answer 30
### Casino
5 chances in 27 throws.
There are 36 different combinations that can be rolled by the casino. Of these, 8 can be won by the player rolling a 2 or a 5, 12 by the player rolling a 3 or a 4, but none if the rolls a 1 or a 6. Therefore of the $6^3$ (216) possible outcomes, the player can win 40, hence a $\frac{40}{216}$ (or 5 in 27) chance of success.

## Answer 31
### Logical Jigsaw

| 7 | 10 | 8 | 11 |
|---|----|---|----|
| 4 | 7  | 5 | 8  |
| 6 | 9  | 7 | 10 |
| 3 | 6  | 4 | 7  |

## Answer 32
### Central symbol
E.

## Answer 33
### Missing hexagon
B.
Reading left to right and up to down, symbols within the first two hexagons are repeated, except where they appear in both, then neither are repeated.

## Answer 34
### Oval numbers
6.
(Top Left x Bottom Right) ÷ (Bottom Left x Top Right) = Middle.

(16 x 7) [112] ÷ (7 x 8) [56] = 2.

(14 x 5) [70] ÷ (2 x 5) [10] = 7.

(18 x 9) [162] ÷ (9 x 3) [27] = 6.

## Answer 35
### Hexagonal numbers
24.
Starting from the top left of each hexagon and moving right, opposite numbers are subtracted, then the three resulting numbers are added together to give the central number.

30 – 20 = 10;

28 – 21 = 7;

14 – 7 = 7;

Total = 24.

**Answer 36**
**Number stack**
D.
As the stacks get smaller the highest number is
omitted and the remaining numbers are reversed.

**Answer 37**
**Grid glitch**
2B.

**Answer 38**
**Symbol hub**
B.

**Answer 39**
**Circular relative**
B.

**Answer 40**
**Sounds spooked**

| SPOOKS | PLACE | STAFF | NOISE |
|---|---|---|---|
| GHOST | MANOR | BUTLER | SCREAM |
| WRAITH | TOWER | COOK | KNOCKING |
| PHANTOM | CASTLE | GAMEKEEPER | GURGLE |
| SPIRIT | ATTIC | MAID | WHISTLE |
| GHOUL | MONASTERY | GARDENER | SHRIEK |

**Answer 41**
**Tile trial**
C.
The figures in the right column and bottom row
contain only those elements which appeared in both
the two previous columns or rows.

**Answer 42**
**Hexagonal quest**
B.
The large, medium and black circles and the triangle
all rotate two angles clockwise and the small circle
rotates one angle anti- (counter) clockwise. Lines
through a shape move in the same direction.

**Answer 43**
**Symbol axis**
A.

**Answer 44**
**Shapescape**

The number of sides to the shape reduces by one
each time.

# Answers 45-54

**Answer 45**
**Deduction sequence**
636.
Each number describes consecutive square numbers, that is:

$1^2 = 1$ (the first square number is 1),

$2^2 = 4$ (the second square number is 4), and so on.

The sixth square number is 36, ($6^2 = 36$) or 636.

**Answer 46**
**Shoe certainty**
70.
If one from each pair is taken, there will be 25 black, 23 white and 21 red shoes – 69 in total. The next one taken, the 70th, must form a pair!

**Answer 47**
**Octagon missing**
E.
Each segment is a mirror image of the segment opposite, but with black and white reversed.

**Answer 48**
**Productivity problem**
2.4%.

**Answer 49**
**Simple square**
A.

**Answer 50**
**Birthday secret**
11.

**Answer 51**
**Number jump**
3.
Working both across and down, the sum of alternate numbers are the same.

**Answer 52**
**Time teaser**
32 minutes or 11.28 a.m.

**Answer 53**
**Circular sequence**
C.
The symbols within two adjacent circles are repeated immediately above, except where they appear in both, then neither do.

**Answer 54**
**Numerical palindrome**
Square the six numbers to produce the following sequence:

961, 9, 25, 121, 529, 169

## Answer 55
### Circular squares
D.
Looking both across and down, the contents of the third square are determined by the contents of the first two squares. Any lines common to the first two squares are not carried forward to the third square.

## Answer 56
### Average cricket problem
58.

$34 \times 11 = 374 + 10 = 384 \div 12 = 32.$

$34 \times 11 = 374 + 58 = 432 \div 12 = 36.$

## Answer 57
### ABC
15599 to 1.
The first card has a 1 in 26 chance of being A, the second a 1 in 25 chance of being B, and the third a 1 in 24 chance of being C. Therefore, the odds are 1 in $(26 \times 25 \times 24) = 1$ in 15600, or 15599 to 1.

## Answer 58
### Symbol point
E.

## Answer 59
### Sneaky sequence
12.
They are consecutive cube numbers: 1, 8, 27, 64, 125, 216, 343, 512 broken down into groups of two digits.

## Answer 60
### Absent number
31.
Starting at 1, jump alternate segments clockwise, double the number each time then add 1.

## Answer 61
### Dotty pentagon
Never.
As can be seen below, at the fifth stage they return to their original positions, completing a loop without ever being together in the same corner.

## Answer 62
### Holiday quintet

| MALE | FEMALE | CITY | TRANSPORT |
|------|--------|------|-----------|
| BOB | MARY | MADRID | BUS |
| SAM | JUDITH | PARIS | FERRY |
| ALF | KATE | NEW YORK | CRUISE |
| LEN | SALLY | ROME | PLANE |
| BEN | FIONA | BERLIN | TRAIN |

## Answer 63
### Circular pyramid
**D.**
Shapes are repeated above from within adjacent circles, but only if the shape appears in both circles.

## Answer 64
### Fractional feature

If you cancel out the shared digits in each fraction, you will produce another fraction of the same value as the original, except for 14/42.

$$\frac{16}{64} = \frac{1}{4}$$

$$\frac{14}{42} = \frac{1}{2} \quad \text{(not 1/3)}$$

$$\frac{19}{95} = \frac{1}{5}$$

$$\frac{26}{65} = \frac{2}{5}$$

## Answer 65
### Bowling dilemma
Both balls go through exactly the same number of revolutions. As the balls are the same size, only the speeds of the revolutions change, not the number.

## Answer 66
### Crazy columns

Column 3 Row 3 = 9.

Column 1 Row 5 = 11.

Column 2 Row 7 = 21.

Column 4 Row 9 = 21.

The sequence of prime numbers 2, 3, 5, 7, 11, 13, 17, 19, 23, 29, 31, 37 alternates between columns 1 and 3. The three times table alternates between columns 2 and 4.

The sequence of square numbers 1, 4, 9, 16, 25 etc. alternates between columns 3 and 1.
The Fibonacci sequence (where each number is the sum of the previous two) 0, 1, 1, 2, 3, 5, 8, 13 etc. alternates between columns 4 and 2.

## Answer 67
### Numerous analogy
**B.**

The second four-digit number and the two digit number are created by joining the sum of the first and third digits and the sum of the second and fourth digits.

6478    6 + 7 = 13,  4 + 8 = 12 (making 1312).

1312    1 + 1 = 2,  3 + 2 = 5 (making 25).

## Answer 68
### Spots before the eyes
The dot marked B. All the other dots are in groups in orbit around dot A.

## Answer 69
### Quarters

## Answer 70
### Russian roulette

It is better to go second. Player 1 takes odd-numbered shots, so dies if there is a first bullet in 1, 3 or 5. The options are 1+2, 2+3, 3+4, 4+5, 5+6, or 6+1. Player 1 dies from 6+1, 1+2, 3+4 or 5+6, 4 of the 6 options; player 2 only dies on 2 of the 6. Much better to be player 2!

## Answer 71
### Pyramidal logic
The missing symbols are:

Reading vertically, alternate lines have the same symbol.

## Answer 72
### Dot dilemma
**A.**
The dot in the top left quarter moves one corner clockwise each stage. The dot in the top right quarter alternates between the same opposite two corners each stage. The dot in the bottom left quarter moves one corner anti- (counter) clockwise each stage. The dot in the bottom right quarter alternates between the same adjacent two corners each stage.

## Answer 73
### Strange sums
**8.**
The figures are derived from Roman numerals split in half.

**9:** ~~IX~~ =4  **11:** ~~XI~~ =6

**12:** ~~XII~~ =7  **13:** ~~XIII~~ =8

## Answer 74
### Dissections
4 possible ways, 5 triangles.

All other possibilities are either rotations or reflections of the above.

## Answer 75
### Missing square
**D.**
Each line across contains one square which contains the combined lines of the other three squares, with the exception that if two lines coincide in the same position in the other three squares, they do not appear in the combined square.

## Answer 76
### Tricky triangles
65.
Square each number and add the results to arrive at the number in the middle.

$5^2 (25) + 6^2 (36) + 2^2 (4) = 65.$

## Answer 77
### Brainstrain

## Answer 78
### Get the needle
3 inches.
The needle goes straight across while the record turns.

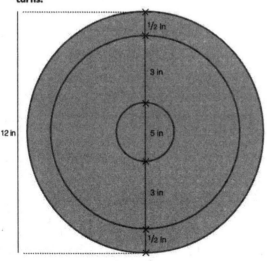

(Not to scale)

## Answer 79
### Bracket gap
84144.
Multiply the three left digits and the three right digits, then express the answers consecutively.

$7 \times 3 \times 4 = 84.$

$2 \times 8 \times 9 = 144.$

## Answer 80
### Pentagonal pyramid
D.
The contents of each pentagon are determined by the contents of the two pentagons below it. Only lines common to both pentagons are carried up to the pentagon above.

## Answer 81
### Multiplication mystery

$4 \times 1738 = 6952.$

$4 \times 1963 = 7852.$

## Answer 82
### Dots spot
A.
Reading both across and down the third square's contents are determined by the contents in the first two squares. Only dots common to the first two squares are carried forward to the third square.

215

# Answers 83-86

## Answer 83
### Twins and triplets
Susan has twins aged 3 and triplets aged 1:

3 x 3 x l x l x l = 3 + 3 + l + l + l.

Anne has triplets aged 2 and twins aged 1:

2 x 2 x 2 x l x l = 2 + 2 + 2 + l + l.

## Answer 84
### Cunning connections

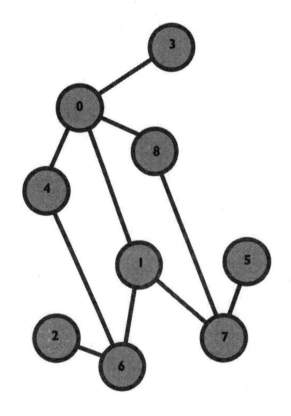

## Answer 85
### Grid symbolism

## Answer 86
### Treasure trail
Visit the squares in the following order:

| 22 | 13 | 18 | 10 | 3 | 26 |
|----|----|----|----|----|----|
| 25 | 30 | 1 | 27 | 21 | 23 |
| 5 | 11 | 28 | 24 | 7 | 2 |
| 12 | 34 | 16 | T | 32 | 14 |
| 19 | 4 | 9 | 6 | 15 | 29 |
| 8 | 17 | 33 | 31 | 35 | 20 |

# Answers 87-95

**Answer 87**
**Teeming with triangles**
23.

**Answer 88**
**Strange journey**
It is impossible! During the first half of the journey I would have used up all the time required in order to travel the whole journey at twice the speed.
Proof: Assume a 100-mile trip. 50 miles at 25 mph = 2 hours. 100 miles at 50 mph = 2 hours.

**Answer 89**
**Griddle**
66.
Each number describes its position in the grid. For example 53 indicates column 5, row 3, and 66 indicates column 6, row 6.

**Answer 90**
**Horse race**

| HORSE | JOCKEY | CAP | SHIRT |
|---|---|---|---|
| ARCTURUS | ARCHIE | BLUE | PLAIN |
| LULLABY | BRIAN | MAUVE | HOOPED |
| MISTRAL | ROGER | RED | STRIPED |
| GENOA | MAURICE | YELLOW | STARS |
| PALISADE | TREVOR | PINK | DIAMONDS |

**Answer 91**
**Melting hex**
B.
Moving clockwise, one side turns into an angle each time. Once a side becomes an angle, it points inward then outward each time (but always starting as an inward angle).

**Answer 92**
**Circles in space**
D.
Reading across and down, repeating symbols appear at the right or bottom.

**Answer 93**
**Hex in space**
B.
Reading left to right and top to bottom, a figure that appears once is repeated.

**Answer 94**
**Shifty symbols**
B.

**Answer 95**
**Natural system**
Base 9.

A, true, 18 = 17, base 10.

B, true, 62 = 56, base 10.

C, untrue, 36 = 33, base 10.

D, untrue, 64 = 58, base 10.

### Answer 96
### Shape shift
A.

They are the top half of the digits 1, 2, 3, 4 etc. on a calculator.

### Answer 97
### Golf poser
They were playing snooker.

### Answer 98
### Full family
Sarah is married to Anthony: the others are all girls.

### Answer 99
### Combination golf

| NAME | COURSE | GOLF CLUB | HOLE |
|------|--------|-----------|------|
| JOE | ST. ANNES | WOOD | 12 |
| BILL | WENTWORTH | DRIVER | 8 |
| FRED | CARNOUSTIE | PUTTER | 14 |
| ALEC | TROON | IRON | 10 |
| COLIN | GLENEAGLES | WEDGE | 6 |

### Answer 100
### Tile teaser
C.
Reading left to right and top to bottom, if an element appears only once, it is repeated right and bottom.

### Answer 101
### Next hex
B.
The pentagon, the Y shape and the circle move two angles clockwise. The T shape moves one angle anti-(counter) clockwise.

### Answer 102
### Slippery symbols
C.

### Answer 103
### Cagey croupiers
19.

### Answer 104
### Hexagons in the round

Horizontally, vertically and diagonally opposite hexagons repeat.

## Answer 105
## Bar queue
E.

The sequence runs:

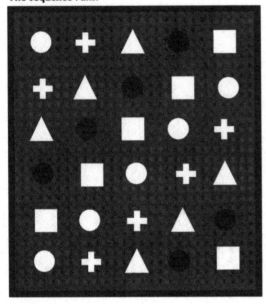

## Answer 106
## Number flags
9.

In each case, the central number is the result of the top two and bottom right numbers being multiplied together, then divided by the bottom left number.

## Answer 107
## Bridge building

| NAME | COUNTRY | TYPE | LENGTH |
|------|---------|------|--------|
| PORT MANN | CANADA | ARCH | 366M |
| SALAZAR | PORTUGAL | SUSPENSION | 1014M |
| GLADESVILLE | AUSTRALIA | CONCRETE | 305M |
| HOWRAY | INDIA | CANTILEVER | 457M |
| COOPER | U.S.A | TRUSS | 488M |

## Answer 108
## Circle of relatives
C.

Black circles turn white and visa-versa, except if its small. Large white circles become small black circles and vice-versa. Squares become triangles, irrespective of size. Small black circles move from top to bottom and vice-versa.

## Answer 109
## Hexagonal roll
E.

The large and small circles connected by a straight line, the line with a cross and the small line with an arrow, move one angle clockwise. The long line with a black dot moves two angles clockwise.

## Answer 110
## Symbolic layers
B.

## Answer 111
## Devilish dice
1.94.

The potential differences between the first dice when thrown showing 1, and the second dice when thrown showing any number from 1 to 6, totals 15. All the other potential differences, (continuing from the first dice showing 2, then 3 and so on) are 11, 9, 9, 11 and 15. Added together these differences total 70. There are 36 potential combinations of the two dice, so 70 ÷ 36 is the average difference: 1.94.

**Answer 112**
**Calculated conundrum**
15189.
Turn the page upside-down, then add.

**Answer 113**
**Box clever**
D.
The black circle rotates clockwise 135˚, the white
circle and the cross both rotate 180˚, and the arrow
rotates 90˚ clockwise.

**Answer 114**
**Town travel**
89.6 m.

**Answer 115**
**Soccer scores**
Player 1 scored 10.
Player 2 scored 12.
Player 3 scored 9.

The incorrect statements were:
Player one, first.
Player two, third.
Player three, third.

**Answer 116**
**Quick detection**
The other two were women.

**Answer 117**
**Cube catch**
C.

**Answer 118**
**Triangular teaser**
7.
In each case, the top two numbers added together,
then divided by the bottom number give the number
in the middle.

**Answer 119**
**Box sympathy**
D.
The dots are in similar circles and the triangle is
below the four linked circles.

**Answer 120**
**Circular spin**
63
A particular combination of the numbers always
gives 9:

$72 + 36 \div 12 = 9.$

$44 + 82 \div 14 = 9.$

$27 + 63 \div 10 = 9.$

**Answer 121**
**Logical segment**
69.
The numbers run clockwise, jumping over adjacent
segments as follows:
$91 - 4 = 87 - 5 = 82 - 6 = 76 - 7 = 69 - 8 = 61.$

**Answer 122**
**Lawn labour**
I hour.
1/4 = 0.250;
1/6 = 0.166;
1/3 = 0.333;
1/4 = 0.250;
Total 1.000 = 1/1 = I hour.

**Answer 123**
**Magic grid**
The numbers have to make a magic grid where each column, row and two diagonals add up to 65.

| 17 | 24 | 1 | 8 | 15 |
|----|----|----|----|----|
| 23 | 5 | 7 | 14 | 16 |
| 4 | 6 | 13 | 20 | 22 |
| 10 | 12 | 19 | 21 | 3 |
| 11 | 18 | 25 | 2 | 9 |

**Answer 124**
**Match magic**

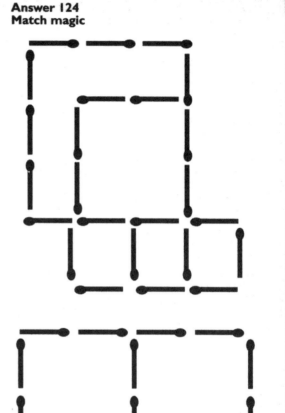

# Answers 125-127

**Answer 125**
**Sharp cards**
We are only interested in the top four cards, which we can call 1, 2, 3, 4. There are only 24 ways in which the four cards can fall. The chances are therefore 1 in 24.

**Answer 126**
**Decimal blunders**

```
  3.76 +        37.6  +
 19.215+       192.15 +
  1.094+   or   10.94 +
 14.7  +       147    +
 38.769        387.69
```

**Answer 127**
**Infinite calculation**
Neither. Infinite sequences like this do not have a finite sum. The current total at any point will always be either 4 or 0, respectively, depending on whether the last part of the sum was +4 or -4.

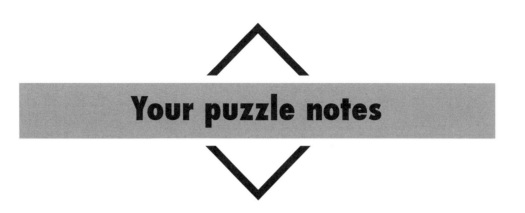

# Your puzzle notes

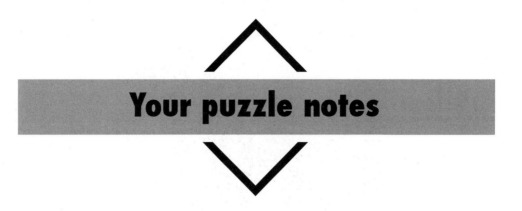

# Your puzzle notes